Henk Haitjema

MECHANICS OF IMMISCIBLE FLUIDS

IN

POROUS MEDIA

by

Arthur T. Corey

WATER RESOURCES PUBLICATIONS

For information and correspondence:
WATER RESOURCES PUBLICATIONS
P.O. Box 2841
Littleton, Colorado 80161, U.S.A.

MECHANICS OF IMMISCIBLE FLUIDS IN POROUS MEDIA

by
Arthur T. Corey

ISBN-0-918334-58-6

U.S. Library of Congress Catalog Number- 86-50589

This publication is printed and bound by Book Crafters, Inc. Chelsea, Michigan, U.S.A.

PREFACE TO SECOND EDITION

The primary objective of this text is the presentation of basic principles of the mechanics of two-phase fluid systems in soils and porous rocks. It is not designed as a reference book or as a thorough review of the literature. Consequently, the scope of the material presented is limited, as explained in the preface to the first edition, to what can be covered in a one-semester course.

The changes appearing in the second edition are made with the objective of providing a clearer presentation of the basic principles. The changes are mostly in Chapters V and VI, which have been rewritten in response to student reactions to the first edition. The word "heterogeneous" has been replaced in the title by the word "immiscible" in order to more clearly reflect the topics discussed.

In particular, the two-phase unsteady flow equation has been written in a different form, and derived in a way that follows more naturally from the preceding material. Moreover, the revised form of this equation is expected to provide a clearer insight into the interrelationship among the various special cases discussed in soil physics and petroleum engineering literature.

An effort is made to make the subject of similarity understandable to students, a majority of whom have previously been exposed only briefly to concepts of dimensional analysis. The approach to the subject in Chapter VI is different from that in most texts dealing with similarity. An attempt is made to provide a less abstract approach than that of dimensional analysis, one which follows more naturally from the students' previous conceptions of geometric similarity.

Thanks are due several colleagues who offered suggestions for the changes incorporated in the revised Chapters V and VI. Special thanks are due Mr. Michael King of the University College, Galway, Ireland who reviewed the entire second edition and contributed many valuable suggestions.

Thanks are also due the author's wife, Vera, for final editing and Miss Sandra Page for typing the manuscript.

November 1985
Fort Collins, Colorado
U.S.A.

Arthur T. Corey
Professor Emeritus
Colorado State University

PREFACE TO FIRST EDITION

This text is concerned primarily with the mechanics of two-phase fluid systems in soils and porous rocks. Applications relating to infiltration, subsurface drainage and production of oil and gas from petroleum reservoirs are included. Each of these applications involves the replacement of one fluid by another, for example, water by air or the reverse and, consequently, are beyond the scope of most texts dealing with groundwater hydrology. Replacement processes, however, are relevant to the overall analysis of groundwater hydrology because they are involved in the interchange between groundwater and water in the atmosphere through the surface mantle. Therefore, the material presented should be of interest to students of hydrology as well as to students of soil science and petroleum reservoir engineering.

The fields of possible applications are extremely diverse. Although the primary sources for material presented in the text are journals of the petroleum industry and soil science, contributions have come from the literature of chemical, agricultural and civil engineers and also from biologists, physicists, applied mathematicians and people working in various industries. An important objective of this text is to prepare students to read literature dealing with replacement processes in porous media regardless of the origin of the literature. It has often happened in the past that scientists in particular fields have ignored the literature generated by others, perhaps because of the inconvenience of digesting unfamiliar notations and somewhat different viewpoints or because of a lack of awareness of the existence of literature on the subject in unfamiliar journals. This text attempts to provide an insight into the relationship among the various viewpoints and an awareness of the extensive literature and broad application for the material presented.

Although reference is made to material from diverse sources, the text does not provide an exhaustive literature review. In the first three chapters particularly, where the emphasis is upon the presentation of basic concepts, little effort is made to present the historical development of the concepts. The motivation for this approach is the observation that students are often distracted by excessive references. In the last three chapters, where concepts are presented which are still undergoing research and change, the sources for the material are given.

The text does not present the broad subject of transport in porous media in all its aspects. The phenomena considered are those that can be analyzed from the viewpoint of traditional fluid mechanics, that is by assuming that each of the two fluid phases constitutes a continuum. The assumptions are made that both fluid phases are Newtonian viscous fluids, that they undergo negligible compression or expansion and that the processes considered are not affected significantly by temperature variations, chemical reactions or changes of phase. Consideration is not given to processes taking place on a "molecular" scale. For example, diffusion which involves molecular mixing and takes place in response to a concentration gradient is not treated. The analysis is based upon the statics and dynamics of fluid particles as the term "particle" is used in fluid mechanics. Consideration is given also to properties pertaining to macroscopic elements of the porous matrix which include solids as well as fluids. In this sense, the subject is treated from both a "microscopic" and a "macroscopic" point of view.

Because of the restrictions upon the material presented, the text is expected to be useful mainly for students in programs that provide other courses relevant to transport in porous media including groundwater hydrology, soil physics and courses treating potential flow theory, the solution of boundary value problems, etc. For students with an appropriate background, it is expected that the material presented can be covered in a one-semester course.

The text is designed primarily for students having a background in elementary fluid mechanics and applied mathematics. Equations are written using the Einstein summation convention which is explained where it is introduced.

Much of the material used in the text is taken from courses taught by the author at Colorado State University during the years 1956 through 1976. The research and ideas of many former graduate students at this University are incorporated in the material presented. The encouragement of the administration and colleagues at the University in the preparation of the text is appreciated. Thanks are due Drs. S. C. Jones of the Marathon Oil Company and E. E. Miller of the University of Wisconsin whose suggestions are incorporated in Chapters V and VI respectively, and P. J. Shuler of the University of Colorado who has reviewed the entire text and made many useful suggestions.

Thanks are also due the author's wife, Vera, for final editing of the entire text and Mrs. Betty D. Hutcheson for typing the manuscript.

Arthur T. Corey
Professor of Agricultural Engineering
Colorado State University

January, 1977
Fort Collins, Colorado
U.S.A.

TABLE OF CONTENTS

TABLE OF CONTENTS - Continued

TABLE OF CONTENTS - Continued

TABLE OF CONTENTS - Continued

TABLE OF CONTENTS - Continued

SYMBOLS

All symbols are defined where they are first introduced. Those symbols which appear in more than one section are listed below along with the section and page where they are first introduced and where the meaning of these symbols is described.

Symbol		Where Introduced Section	Page
$D(\theta)$	A coefficient used in the diffusion form of the Richards equation	3.9	111
d	pore size	1.3.3	5
$\bar{d}$	average pore size	1.3.3	5
e	a subscript meaning effective		
$\underline{e}$	a unit vector		
F_w	fractional flow function for wetting phase	5.3.1	201
F_{nw}	fractional flow function for non-wetting phase ($F_{nw} = 1-F_w$)	5.3.2	203
f_w	$f_w = 1 - f_{nw}$		
f_{nw}	a function of permeability and viscosity ratios	5.3.2	203
F.C.	field capacity	2.3.4	42
g	(1) magnitude of force per unit volume due to gravity (2) a subscript referring to a gaseous phase	2.1.1	34
$\underline{g}$	force per unit volume due to gravity (a vector)	2.1.1	27
H	piezometric head	3.2.2	80
H_k	effective permeable height	4.4	155
h	elevation above a datum	2.2	34
h_c	capillary pressure head	4.2.1	135

h_d	elevation in a soil-water system at which $p_c = p_d$, or capillary pressure head equal to $p_d/\Delta\rho g$	2.5.1 4.2.1	59 135
i,j,k	subscripts referring to orthogonal directions	2.1.1	29
K	hydraulic conductivity	3.8	100
K_m	maximum field conductivity	4.4	155
k	permeability (intrinsic permeability	3.7	97
k_g	effective permeability to gas	5.3.2	205
k_{nw}	effective permeability to non-wetting phase	3.7	97
k_o	effective permeability to oil	5.3.2	205
k_{rw}	relative permeability to wetting phase	3.7	97
k_{rnw}	relative permeability to non-wetting phase	3.7	97
k_s	dimensionless shape factor	3.3.4	86
k_w	effective permeability to wetting phase	3.7	97
nw	subscript referring to non-wetting phase	1.4.4	12
m	subscript indicating critical or maximum value	2.3.1	39
p	fluid pressure	2.1.1	28
p_b	bubbling pressure	2.3.2	39
p_c	capillary pressure	1.5	13
p_d	displacement pressure	2.3.2	39
p_e	entry pressure	2.3.2	40
p_{nw}	pressure of non-wetting fluid	1.5	13

p_w	pressure of wetting fluid	1.5	13
p^*	piezometric pressure	3.2.3	80
Q_i	cumulative inflow	5.3.1	202
Q_o	cumulative outflow of oil	5.3.2	202
Q_w	cumulative outflow of water	5.3.2	202
q	volume flux rate	3.7	96
q_{nw}	volume flux rate of non-wetting phase	4.1.2	131
q_t	total flux rate equal to $q_w + q_{nw}$	5.3.1	200
q_w	volume flux rate of wetting phase	3.7	96
R	hydraulic radius	3.4	86
r	(1) radius, or (2) subscript indicating a residual saturation	1.5.1 2.3.3	15 40
S	saturation	1.4.4	11
S_e	effective saturation	2.3.3	41
S_r	residual saturation	2.3.3	40
S_y	specific yield	2.5.1	59
s	specific surface	1.3.2	4
$\underline{s}$	displacement vector	3.2	78
T	tortuosity	3.5	92
t	time		
u	velocity of center of mass of fluid element	3.1.1	72
V_m	reference volume of porous medium	1.3.1	3
V_p	volume of interconnected pore space in a sample or in a reference volume	1.3.1	3

V_s	volume of bulk sample	1.3.1	3
W	liquid content by weight	1.4.4	11
w	a subscript indicating a wetting phase	1.4.4	12
wp	wetting perimeter	2.4	48
i,j,k	coordinates in i,j,k directions respectively	2.1.1	29
α	contact angle	1.5.1	15
γ_s	specific gravity of solid particles	1.4.4	11
Δ	difference (increment)		
ε	empirical exponent	3.6.1	95
η	empirical exponent	3.8.6	109
θ	volumetric water content	1.4.4	11
λ	empirical exponent used as index of pore-size distribution	2.3.6	46
μ	dynamic fluid viscosity	3.1.3	74
ρ	fluid density	2.1.1	27
σ	interfacial force	1.5.1	14
τ	intensity of shear	3.1.3	73
Φ	force potential	3.2.1	72
ϕ	porosity	1.3.1	2
ϕ_e	effective porosity	2.3.3	41
ψ	suction or negative pressure head	1.5	13
ω	angular velocity	2.5.4	65
$\hat{}$	symbol indicating a scaled variable	4.2.2	138

CHAPTER 1

PROPERTIES OF POROUS MEDIA AND FLUID MIXTURES

1.1 POROUS MEDIA

In the most general sense of the word, the term "porous" could be applied to all matter, because all matter contains non-solid space. However, for the purpose of this text, additional restrictions are placed upon matter which is considered porous. These are:

(1) The non-solid space within the solid matrix is interconnected.
(2) The smallest dimension of the non-solid space must be large enough to contain <u>fluid particles</u>; that is, it must be large compared to the mean-free path of fluid molecules.
(3) The dimensions of the non-solid space must be small enough so that when interfaces between two fluids occur within the non-solid space, the orientation of interfaces will be controlled largely by interfacial forces.

The first restriction eliminates consideration of a solid having only isolated pockets of non-solid space. It also eliminates consideration of a bundle of capillary tubes which are not cross-connected. A single capillary tube might be regarded as a porous medium, but not a bundle. The second restriction eliminates consideration of molecular transport through solids with non-solid spaces of such small dimensions that true convection cannot occur and fluid mechanics cannot be applied. The third restriction eliminates consideration of a network of pipes.

As an explanation of the latter point, consider the distribution of a mixture of water and air in a pipe as compared to that in a capillary tube, as illustrated in Figure 1-1.

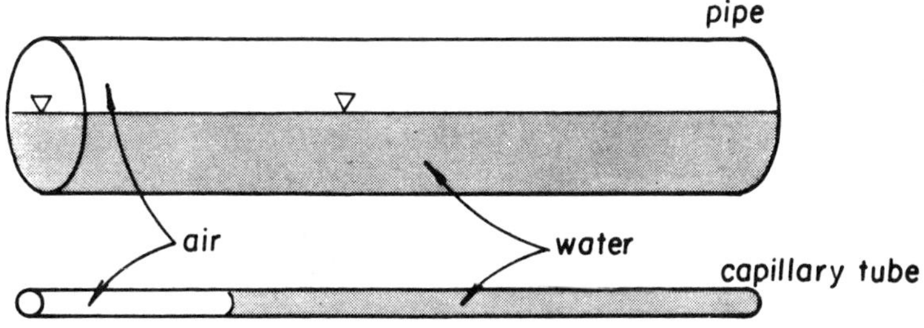

Figure 1-1. Orientation of interface in pipe as compared to that in capillary tube.

1.2 TYPES AND OCCURRENCE

Primary concern in this text is with flow in earth materials, e.g., soils and porous rocks. Most soils and porous rocks consist of a class of porous materials which are granular in nature. That is, they are composed of relatively solid grains with non-solid space enclosed. The non-solid space is called pore space. The grains are sometimes cemented at points of contact with a variety of cementing agents, in which case they are said to be consolidated. A sandstone is an example of a consolidated porous medium. In other cases the grains are not cemented at points of contact and such materials are said to be unconsolidated. Most soils and sands are examples of this type.

Another type of porous rock consists of non-solid space which was created by the evolution of gases during crystallization and another by the subsequent solution of soluble constituents in water. Examples of the latter type are vugular limestones and dolomites. The pore space in such rocks consists of channels called vugs formed by solution. Not all porous limestones are vugular; however, some consist of cemented fragments of more or less solid limestone. The latter are called intergranular limestones or intergranular dolomites.

Other types of porous material important in various fields include wood, living plant and animal tissue, textiles, building materials, filter materials and others.

1.3 CHARACTERIZATION OF PORE SPACE

The hydraulic behavior of fluids in porous media depends to a large degree upon the geometry of the non-solid space enclosed. In the case of granular material, the geometry is influenced by the size and shape of individual grains and how they are cemented together.

Some granular media are said to possess structure. This implies that individual grains are cemented together to form porous aggregates which in turn enclose additional pore space. Pore space enclosed within individual grains is called primary pore space and that enclosed between the aggregates is called secondary pore space. A granular medium with structure contains a larger total volume of pore space in relation to its mass than one which does not possess structure.

Important characteristics of porous materials in relation to hydraulic behavior are described below.

1.3.1 Porosity - The average porosity of a sample of porous medium is defined as the ratio of interconnected pore volume V_p to the total volume V_s of the sample. The volume

V_s includes solid as well as non-solid space. In some literature, average porosity may be called simply porosity and designated by ϕ.

Sometimes, however, ϕ refers to a property which is assumed to apply to a point and which can vary in space. In this case, ϕ is defined as the ratio V_p/V_m in which V_p is the volume of pore space enclosed in V_m, a small volume of porous medium containing the point under consideration. To serve as a reference volume for this definition, V_m must be small (compared to the entire system) but large compared to individual solid grains. It must be sufficiently large that the ratio V_p/V_m will not change abruptly if a slightly larger reference element (containing the same point) is considered, unless the larger element encompasses a boundary between different types of media.

In some analyses, ϕ is treated as if the ratio V_p/V_m is equal to the derivative of V_p with respect to V_m. Clearly an actual derivative is equal to either 0 or 1 depending upon whether the point under consideration is located within a solid grain or within the pore space, but this is not what is meant by ϕ. Consequently, the interpretation of ϕ as the derivative of V_p with respect to V_m requires a special definition of the derivative as the limit of V_p/V_m as V_m approaches a critical size somewhat larger than that at which the ratio V_p/V_m undergoes abrupt changes as V_m changes. Abrupt changes in the ratio will occur, for example, if V_m is of a size of the same order of magnitude as that of individual grains.

Porosity at a point, therefore, is an abstraction that cannot be measured experimentally. Only the average porosity of a sample of rock or soil can be measured. This may be accomplished in one of several ways, some of which have been described by Collins (1961). Some of the methods determine the volume of liquid required to fill the pore space. Others determine the volume of the solid and this is subtracted from the total volume to obtain the pore volume. Still others utilize Boyles' law to compute the pore volume after allowing gas in the pores to expand. Factors affecting porosity include:

(1) Structure - Media with structure have larger porosities than media without structure.

(2) Shape of grains - A medium consisting of flat platelets can be packed so that the porosity is much smaller than a collection of spheres. However, it

would be possible to stack the platelets so that the porosity would be much greater than that for spheres.

(3) Grain-size distribution - A medium consisting of spheres of varying sizes will normally have a smaller porosity than one consisting of spheres of a single size. The smaller spheres may fit into spaces between larger spheres, thus reducing the porosity.

(4) Mixing - A medium consisting of two different sizes of spheres, for example, but with the two sizes segregated into different regions will have the same porosity as a single-size medium. If the two sizes are mixed, the porosity will be reduced.

(5) Packing - The way that individual grains are arranged can be influenced by the mechanical conditions at the time of packing, for example, whether the particles settled out of water or were deposited by wind, or were deposited by some other geological process. Laboratory samples can be affected in this regard by the mechanical manipulation used in packing. The effect can be visualized by considering a stack of cards and visualizing the non-solid space enclosed when they are in a usual deck as compared to that which would be enclosed if they were arranged in a cubicle pattern.

(6) Cementation - The volume of cementing material, which may have precipitated from solution after the particles were deposited, reduces the porosity. Consequently, consolidated sandstones typically have smaller porosities than unconsolidated sand deposits.

Approximate porosities that could be expected in soils and porous rocks under various conditions include:

Consolidated sanstones	0.10-0.30
Uniform spheres packed to theoretical minimum porosity	0.26
Uniform spheres with normal packing	0.35
Unconsolidated sands with normal packing	0.39-0.41
Soils with structure	0.45-0.55

1.3.2 Specific surface - The ratio of internal solid surface area to the total volume is called specific surface and is designated by s. In some texts the ratio of surface area to the mass of solid matrix is called specific surface. Specific surface may also be regarded as existing at a point, employing a concept analogous to that used for porosity.

In this case, the ratio of internal solid surface area to the total volume of a small element of the medium is defined as the specific surface of a point contained in the element. The volume element to be considered must be small relative to the volume of the system being analyzed and large enough that a slight increase in the volume would not produce a significant change in the ratio evaluated. In contrast to porosity, which is dimensionless, specific surface has the dimension of L^{-1}.

As in the case for porosity, specific surface is not something that can be determined experimentally at a point. A method of determining specific surface for a porous medium sample has been described by Bower and Goertzen (1959). In this case, a dry sample is allowed to adsorb ethylene glycol molecules from a saturated solution until a monolayer of these molecules has formed on the solid surface. From the gain in weight of the sample produced by the monolayer, the surface area of the solid is computed. A method of determining specific surface by the adsorption of nitrogen gas has been described by Donaldson et al. (1975).

In the case of granular material consisting of grains of relatively uniform size, the specific surface may be estimated from hydraulic measurements which are described in Chapter III.

Specific surface is affected by the size and shape of individual grains, and to a lesser extent by structure. The smaller and flatter the particles, the greater the specific surface. Clay (which belongs to a class of minerals known as hydrous alumina silicates and consists of tiny plate-like crystals) has an enormous specific surface compared to other earth materials. Consequently, by far the most important factor (in determining the specific surface of earth materials) is the amount and types of clay contained.

Three important types of clay are recognized. These differ somewhat in chemical composition and there are many subtypes differing in respect to crystalline form. The three main types are: kaolinite, montmorillonite, and illite. Of the three, montmorillonite has the largest specific surface, illite the next largest and kaolinite the smallest. Expressed as square meters per gram, typical values are:

montmorillonite	800
illite	175
kaolinite	45

1.3.3 Pore size - An average pore size $\bar{d}$ for a porous medium is defined as the ratio ϕ/s. The use of the term "average", in this case, does not mean that an entire sample of porous medium is considered. In fact $\bar{d}$ may be conceptualized, but not measured, as a property applicable at a point since both ϕ and s may be regarded as applicable at a point.

The symbol $\bar{d}$ is used to distinguish this parameter from pore size d which is defined in an analogous way in respect to a smaller reference volume, one that is of the same order of magnitude (or smaller) than that of individual solid grains. Clearly, d varies in space depending upon which element of space is considered. It varies from point to point even within an element the size of that required for the definitions of ϕ and s.

Pore size has the dimension of length and is physically analogous to hydraulic radius as the latter term is used in hydraulics. To visualize this analogy, consider a section of a capillary tube and note that the cross-sectional area of the tube divided by its internal perimeter is equal to the ratio of its volume to its internal solid surface area, provided that the cross-sectional area is uniform. If the cross-sectional area is not uniform, the ratio of volume to surface area gives the "average" hydraulic radius of the section.

Clearly, pore size is associated with grain size and grain-size distribution. The smaller the grains the smaller is the pore size. However, materials with structure may have some large pore sizes associated with the secondary pore space, even though the primary pore space is characterized by a small pore size.

1.3.4 Pore-size distribution - Distribution of pore sizes is a concept applicable to d, but not to $\bar{d}$. The fraction of pore space represented by various ranges of d (within a volume element) has an effect of equal importance to that of $\bar{d}$. The distribution of d cannot be measured directly, but there is a way of defining and measuring an index of pore-size distribution. This is described in Chapter II where the concept of pore-size distribution and its effect on fluid flow is discussed in detail.

To some extent pore-size distribution is related to grain-size distribution. A wider distribution of grain sizes result in a wider distribution of pore sizes, other things being equal. Thorough mixing, however, may produce a relatively uniform pore size even with a wide distribution of grain sizes.

A very wide range of pore sizes is obtained in granular material only if it possesses structure.

1.3.5 Stability and other characteristics - The degree to which the characteristics described above remain constant in the presence of fluids (and under mechanical forces) is called stability. Both mechanical and chemical stability are important characteristics of porous media.

In respect to earth materials, by far the most important stability factor is the effect on pore-size distribution produced by the sensitivity of clay crystals to water. Clay crystals are usually stacked together like cards in a deck, the individual plates being bound together tightly in some cases and less tightly in others. The effect of water is to spread the plates, a process called clay swelling or dispersion. The opposite process is called shrinking or flocculation.

When clay is dispersed, the effect is to reduce the larger pore sizes and thus decrease the range of the pore-size distribution. This has an enormous effect on the hydraulic properties of the medium which is explained in Chapter III.

Of the three types of clay, montmorillonite is the most sensitive to water solutions, kaolinite is least sensitive, and illite is intermediate. Note that the order of sensitivity parallels that for the specific surface of the three types.

Water has less tendency to enter the space between clay plates (and spread them apart) when the solution is highly concentrated with electrolytes. Furthermore, certain ions inhibit swelling and dispersion much more than others. There is a tendency for ions of highest positive valence to be most effective. Thus, Al^{+++} is more effective than Ca^{++} and Ca^{++} is more effective than Na^{+} or H^{+}. Potassium, K^{+}, is somewhat of an exception in that because of its particular ionic size, it inhibits swelling more than other monovalent ions.

In general, earth materials (rocks or soils) are more stable in the presence of hydrocarbon liquids than in the presence of water, especially if they contain significant quantities of clay or organic matter. The reason is that hydrocarbons have much less tendency to produce clay swelling. This is another reason why oils are often used in laboratories for flow experiments.

Mechanical manipulation, especially when wet, of unconsolidated materials also has an effect on the stability of unconsolidated media in respect to pore-size distribution. The tendency is for such manipulation to break down aggregates and reduce the larger pore sizes.

The shape of pore space and the degree to which it is interconnected are also important.

Homogeneity and isotropy are important properties, but these are best characterized in respect to flow behavior and are defined in that context in Chapter III.

1.4 FLUIDS IN POROUS MEDIA

Fluids belong to a class of matter the boundaries of which depend upon the geometry of the solid within which it is enclosed. This dependence results from the inability of fluids to support shearing stresses without continuous deformation.

The magnitude of shear stresses in fluids is a function of the rate of deformation of fluid elements. In some fluids, the magnitude of shear stress is approximately linearly related to the rate of deformation. Such fluids are called Newtonian viscous fluids and this text deals almost exclusively with these. Air, water and most petroleum fluids can usually be regarded as Newtonian without serious error.

Although the text is restricted to a consideration of Newtonian viscous fluids, mixtures of fluids such as air and water, gas and oil or water and oil are considered.

1.4.1 Concept of a continuum - All fluids consist of particles such as molecules or ions, but this aspect of fluids is ignored in fluid mechanics by an artifice known as the continuum concept. Neither the properties nor motion of individual molecules are described by fluid mechanics.

The word "continuum" is related to the purely mathematical concept of continuous functions. A function $f(x)$ is said to be continuous at a point $x = x_o$ if $\lim_{x \to x_o} f(x) = f(x_o)$ regardless of how x approaches x_o. The foregoing analytical definition of continuity is a formulation in mathematical language of the intuitive concept of continuity; that is, if the function $f(x)$ is represented by a graph, and if it is continuous, the graph will have no breaks in the interval within which the function is continuous.

In fluid mechanics, the pertinent functions are the relationships between the space coordinates and the variables: pressure, velocity, and density.

If the fluid is to be regarded as a continuum, it must be possible to define each of these variables (at every point within the region under consideration) as a property of a very small element containing the point. In addition, it must be possible to assume that derivatives with respect to the space coordinates (at least the first and second derivatives) exist at every point. With some exceptions, fluid mechanics also assumes that pressure and velocity are differentiable to any order. In other words, it is usually assumed that these variables are analytic as well as continuous functions of the space coordinates.

One physical implication is that it theoretically would be possible to divide the fluid volume indefinitely without changing its basic character. This is obviously not realistic because all fluids consist of molecules between which there is empty space. The definitions of density, velocity, and pressure clearly are meaningless in respect to volume elements located in spaces between molecules.

The unrealistic assumption upon which fluid mechanics (a branch of continuum mechanics) is based, does not lead, necessarily, to significant errors in its application. For most cases in which fluid mechanics is applied, the assumption that the fluid is a continuum is entirely adequate. These are cases in which the dimensions of the fluid system under consideration are very large compared to the average distance between individual fluid molecules.

In porous media, however, it is not always possible to define pressure, density and velocity as properties of a small volume containing a point, without the reference volume approaching the cube of the pore size. When the fluid under consideration is gas at atmospheric pressure, for example, the use of a continuum analysis leads to large errors if the media are "fine-grained".

The continuum assumption is a great convenience because it permits a description of the physical behavior with differential expressions. Without such an assumption, the analyses (ordinarily performed by the methods of fluid mechanics) would become very cumbersome.

It is obvious, however, that the concept of a continuum (in the sense that it is used in fluid mechanics) cannot be extended to immiscible fluids separated by interfaces across which pressure discontinuities exist. In the latter case it is necessary to consider each individual fluid phase as a separate continuum.

1.4.2 Fluid elements - In fluid mechanics the physical variables (including velocity) are specified at points within the fluid system. As previously stated, the variables are defined in respect to small volumes of fluid containing the points under consideration. A reference volume must be small compared to the pore size but it must not be smaller than a limiting size which can be understood by referring to Figure 1-2.

Figure 1-2 shows a plot of hypothetical data that might be obtained by determining the density of volume elements of a fluid in the form of small cubes. If the dimensions of the elements were of the same order as the mean free path of the fluid molecules (or smaller), the density would vary erratically and be greatly different from one instant to another. The density variation would result from the random motion of the molecules and would depend on the number of molecules that

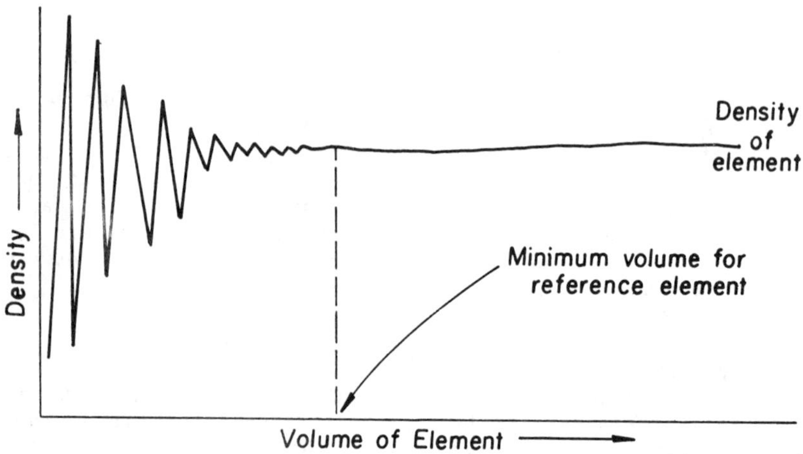

Figure 1-2. Density as a function of element volume.

happened to be in a particular volume element at a given instant. As the sample size is increased, the variation would decrease. At some critical dimension, large compared to the mean free path, the variation will be negligible and a small increase in size of the element will not change the measured density significantly.

An element at least as large as the critical size, but small compared to the fluid system as a whole, is sometimes called a fluid "particle". The critical size is, of course, much larger for gases than for liquids. No practical problems involving liquids are likely to be encountered for which the size limitation of a fluid particle will invalidate a continuum assumption. For gas flows, on the other hand, many engineering problems can not be analyzed using the continuum assumption without introducing significant error. An example of the latter is the flow of gas in fine-grained porous media at atmospheric pressure.

1.4.3 Two-phase fluid systems - When two or more fluids exist within the pore space, they are separated by boundaries (called interfaces) across which discontinuities in density and pressure exist. The existence of interfaces is characteristic of what are called immiscible fluids. In the case of miscible fluids, there is no distinct boundary, at least on a microscopic scale. An example of the latter situation is a groundwater aquifer into which salt water has intruded. On a macroscopic scale it is possible to find distinct regions of salt and fresh water, but on a microscopic scale there is no interface as this term is used here.

In this text, the concept of separate fluid phases implies that interfaces exist. An interface is made possible by the existence of forces (called interfacial forces) that act only at boundaries between separate phases and are tangential to the boundaries. When the boundaries are curved, as they usually are in porous media, the tangential interfacial force produces pressure discontinuities at the interfaces.

Each phase may consist of a number of chemical constituents, but they are assumed to be homogeneous within themselves and to constitute physical continua to which the mathematical methods of fluid mechanics can be applied. Fluid mechanics (in the sense this term is used here) cannot be applied to the mixture of phases.

1.4.4 Liquid content - When a porous medium is occupied by more than one fluid phase, there are several ways commonly used to describe the amount of each fluid contained in the medium.

For example, in the case of a two-phase system such as a liquid and gas, the volume fraction of the total pore volume (of a medium element) that is occupied by the liquid is called saturation and is designated by the symbol S. Note that S can be conceptualized (but not measured) as a property applicable at a point and varying in space in a manner entirely analogous to that used in defining porosity at a point. The parameter S should not be confused with the term "saturated" which means that only a liquid phase exists, or in other words, S is 1.0.

Another way of expressing liquid content is as a volume fraction of the volume of porous medium in which it is contained. The latter parameter is called volumetric water content by soil physicists and is often designated by θ. A third method is as a fraction W of the dry weight of porous solid in which the liquid is contained. Designation of liquid content by the parameter W is found most often in the literature of agronomists and soils engineers.

The relationship among the three expressions for liquid content is

$$\theta = S\phi = \gamma_s(1-\phi)W \quad , \qquad 1.1$$

γ_s being the specific gravity of the solid. The question of which of these expressions for liquid content is most convenient depends upon the particular application under consideration. In this text, each of them is used where appropriate.

Petroleum scientists frequently must consider systems in which three fluid phases are contained in the pore space of porous rocks such as brine, oil and gas. In their literature, the symbols S_w, S_o and S_g may designate saturations of brine, oil and gas, respectively, each representing a fraction of the total pore volume of the reference element.

Whenever two or more fluid phases occupy a porous medium, it is inevitable that one of the fluids will be adsorbed on the solid surfaces more strongly than the other fluid. The fluid which is most strongly adsorbed (and which displaces the other fluid from the adsorbed film) is called the wetting fluid or wetting phase. The displaced fluid is the non-wetting phase. When referring to the wetting phase, the fluid properties such as pressure, density, or velocity are designated as p_w, ρ_w and u_w, etc. Similarly, the subscript nw designates the properties of the non-wetting phase.

In most cases, liquids are adsorbed more strongly than gases and as a consequence, in a two-phase system involving a liquid and gas, the liquid usually will be the wetting phase. An exception is the case of a mercury-gas system for which mercury is the non-wetting phase. If the term "saturation" is used without adjectives to describe fluid content, it usually refers to a volume fraction of the wetting phase in a two-phase system. The content of other phases is usually referred to as "gas saturation" or "non-wetting phase saturation", etc.

1.5 CAPILLARITY

At boundaries between phases, forces of cohesion between fluid molecules are not (within themselves) balanced. There is a force component from unbalanced cohesion which acts tangentially to the boundary and which is called interfacial force. Discussions of the origin of interfacial force have been presented by Adam (1968) and the Encyclopedia Britannica (1964), among others.

Interfacial force acts in a direction tending to contract the interfacial area, in a manner somewhat analogous to tension in a stretched membrane. For this reason, it is often called surface tension, the dimensions being force per unit length or energy per unit area. The analogy between interfacial force and the tension in a stretched membrane is not complete, however, because in the case of the former, the force is unrelated to deformation. Interfacial force is a function only of the physical and chemical properties and state of the two fluid phases in contact.

The failure (in some respects) of the analogy between interfacial force and tension in a membrane has led some authors to reject the concept of interfacial force as a physical reality. They prefer to speak only of interfacial energy. This view does not seem to be tenable, however, considering the fact that energy can only be defined in terms of force fields, and if there is no unique force at points in the interface, there could be no unique energy associated with the interface.

The resultant of interfacial force acting on a curved interface is balanced, at equilibrium, by a difference in pressure at points of contact between fluid phases. Without interfacial force, which results in the difference in pressure, separate phases would not exist; that is, distinct interfaces would not exist. The difference in pressure, called capillary pressure, is designated by p_c and is defined by

$$p_c = p_{nw} - p_w \quad , \qquad 1.2$$

p_{nw} and p_w being the pressures of the non-wetting and wetting phases respectively. In soils literature, p_c is sometimes called "suction" or "matric suction" and designated by ψ. However, the term suction and the symbol ψ are sometimes used to indicate the "negative pressure head" of water in a water-air system, it being assumed that the air is everywhere at zero gauge pressure. In the latter case, ψ is $p_c/\rho_w g$ which may also be called soil water tension. In any case, suction, capillary pressure and tension are more-or-less equivalent concepts.

One effect of interfacial force is a tendency to compress the non-wetting phase relative to the wetting phase. For example, in a water-gas system, water tends to be preferentially adsorbed by the solid surfaces and the gas is compressed if it is entirely surrounded by water. The gas pressure, therefore, is higher than that of the water. In a water-oil-gas system, the gas will usually be the highest pressure, the oil at the next highest and water at the lowest pressure when equilibrium exists.

In a system in contact with the atmosphere, such as a soil, either the wetting phase (water) or the non-wetting phase (air) will normally be at atmospheric pressure. If the soil is sufficiently desaturated for an interconnected air phase to exist, the air will be at atmospheric pressure and the liquid will be at less than atmospheric pressure. This is the reason that soil scientists refer to water as being under "suction" or "tension". On the other hand, if the soil is in contact with water at atmospheric pressure, that is, if it is flooded, some air may be entrapped within the pore space which is below the level where water is at atmospheric pressure, and this air will be at a pressure greater than atmospheric.

1.5.1 Factors affecting capillary pressure - The difference in pressure between phases occupying the pore space of a porous medium is related to gravity, saturation, pore-size, pore-shape, interfacial forces, the angle at which fluid-fluid interfaces contact solid surfaces, the density difference between phases and the radii of curvature of interfaces. These factors are not all independent variables in respect to their affect on p_c. There is, however, an interrelation among them.

The way that gravity and saturation are related to p_c is discussed in Chapter II which deals with statics of fluids in porous media. The relationship among p_c and the other variables mentioned above can be visualized by considering the following analysis in reference to Figures 1-3, 1-4 and 1-5.

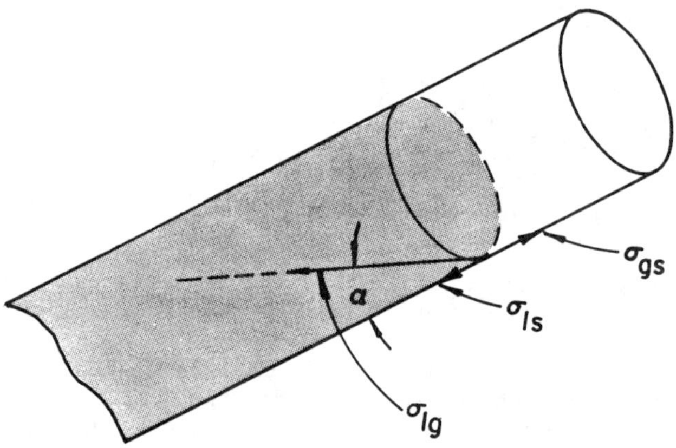

Figure 1-3. Liquid-air interface across a capillary tube.

First, a balance of forces is considered on a free body consisting only of the line of contact of the fluid-fluid interface with the solid (see Figure 1-3). There may be three interfacial forces acting on the line of contact. These are:

(1) liquid-solid $\sigma_{\ell s}$,

(2) gas-solid σ_{gs},

(3) liquid-gas $\sigma_{\ell g}$.

These forces act at all points along the line of contact in a direction indicated in Figure 1-3. Note, however, that gravity and pressure forces do not act on this free body because it has neither mass nor area. Also, the component of $\sigma_{\ell g}$ which is normal to the tube surface is balanced by the normal force exerted by the solid on the line of contact. The balance of forces, therefore, is given by

$$2\pi r_t[\sigma_{gs} - \sigma_{\ell s} - \sigma_{\ell g}(\cos\alpha)] = 0 \qquad 1.3$$

where r_t is the radius of the tube.

Equation 1.3 is useful to show, in a qualitative way only, how the contact angle α is related to the three interfacial forces. Cos α cannot actually be computed from this because σ_{gs} and $\sigma_{\ell s}$ cannot be measured. It is instructive to note, however, that $\cos\alpha$ must adjust so that the component of $\sigma_{\ell g}$ (in the direction of the axis of the tube) is equal to the difference $\sigma_{gs}-\sigma_{\ell s}$.

Next, a balance of forces is considered on a free body consisting of the interface as a whole. In this case, the pressure difference p_c across the interface must be considered but gravity does not, because the interface has area but no mass. The unknown interfacial forces, σ_{gs} and $\sigma_{\ell s}$, can be eliminated from consideration by taking as a free body that part of the interface which is not in direct contact with the solid. The balance of forces is further simplified by considering only the components of the pressure forces normal to a plane passing through the line of contact of the interface with the solid.

With these simplifications, the balance of forces is given by

$$\bar{p}_c(\pi r_t^2) = \sigma_{\ell g}\cos\alpha(2\pi r_t)$$

or

$$\bar{p}_c = \frac{2\sigma}{r_t}\cos\alpha \qquad 1.4$$

where $\bar{p}_c$ is the average capillary pressure over the interface and σ is the fluid-fluid interfacial force previously designated by $\sigma_{\ell g}$.

Equation 1.4 shows that $\bar{p}_c$ varies inversely with the radius of the tube across which it is positioned and directly with the cosine of the angle of contact. However, capillary pressure p_c varies from point to point over the interface because of gravity. This variation is associated with a corresponding variation in curvatuve of the interface with elevation over the interface.

The dependence of p_c on curvature is analyzed with reference to Figure 1-4.

Figure 1-4 represents a small segment of a curved interface containing the point p. The point is at the center of the segment which is approximately square. The edges of the segment are each of length ℓ. The angles θ_1 and θ_2 are those subtended by an arc length of $\ell/2$ in orthogonal planes normal to the segment at p, with radii of curvature r_1 and r_2 respectively.

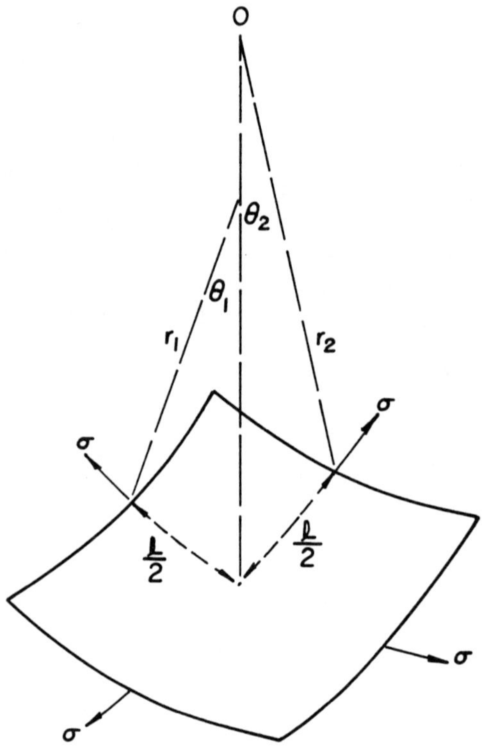

Figure 1-4. Relationship between p_c and curvature at a point.

For small θ,

$$\sin\theta_1 \cong \frac{\ell}{2r_1}$$

and

$$\sin \theta_2 \cong \frac{\ell}{2r_2} \quad .$$

Balancing the force components (pressure and interfacial) normal to the segment at p results in

$$p_c \ell^2 \cong 2\ell\sigma(\sin \theta_1 + \sin \theta_2)$$

or

$$p_c \cong \sigma(\frac{1}{r_1} + \frac{1}{r_2}) \quad .$$

In the limit, as $\ell \to 0$, the relationship becomes exact; that is,

$$p_c = \sigma(\frac{1}{r_1} + \frac{1}{r_2}) \quad . \qquad 1.5$$

Equation 1.5 is usually known as the LaPlace equation of capillarity. It is, probably, the most important equation relating to the behavior of heterogeneous fluids in porous media.

Note that Equation 1.5 evaluates the pressure difference across the interface at a point. The capillary pressure is given as a function of σ and curvature at the point only. The contact angle and pore size are not involved explicitly.

It is informative to consider the relationship between Equation 1.4 and 1.5 by applying Equation 1.5 to an interface across a tube as illustrated in Figure 1-3. For a very small tube such that fluid weight has negligible effect, p_c is constant over the interface which is a segment of a sphere. For example, with a zero contact angle, the interface is a hemisphere with radii of curvature equal to the radius of the tube. Since for this case r_1 and r_2 are equal, Equation 1.5 reduces to the same form as Equation 1.4. If the angle of contact is not zero, equating Equation 1.4 to Equation 1.5 shows that the radius of curvature of the interface is the radius of the tube divided by cos α. It is emphasized, however, that this result assumes the absence of a gravitational effect.

The quantity $(1/r_1 + 1/r_2)$ is sometimes referred to as the mean curvature of an interface at a point. Mean curvature is affected by pore shape as well as pore size. The effect of shape can be visualized by considering an interface across a pore space with a shape that is different from that of the tube illustrated in Figure 1-3. In the case under consideration, shown in Figure 1-5, the interface is positioned across the space between two flat parallel plates.

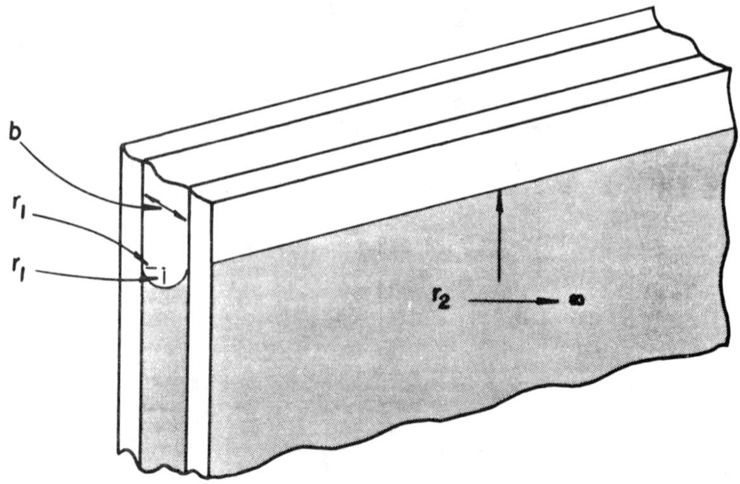

Figure 1-5. Interface across space between parallel flat plates.

The tendency is for an interface to reach (at equilibrium) an orientation, position and shape such that the total potential energy of the system (relative to the force fields of gravity, interfacial energy and pressure) is a minimum. In the case under consideration, this tendency results in an interface which is horizontal in vertical planes parallel to the plates; that is, $r_2 \rightarrow \infty$. This is produced by the dominant effect of gravity in controlling the curvature in parallel planes, the dimension of the fluid body being large in this direction. However, in vertical planes normal to the plates, if the spacing "b" is small, r_1 is $b/(2 \cos \alpha)$. For this case, therefore, Equation 1.5 reduces to

$$p_c = \frac{2\sigma \cos \alpha}{b} \quad . \qquad 1.6$$

If the dimension of the pore space parallel to the plates is relatively large, say more than a few millimeters, the magnitude of p_c is unaffected by further increases in this dimension. Capillary pressure in such cases is controlled only by the small dimension normal to the plates.

In the case of interfaces in pore spaces enclosed in granular porous solids, the orientation of interfaces is such that the area of interface tends to be a minimum consistent with the fraction of wetting or non-wetting fluid contained in the space.

1.5.2 Factors affecting surface tension

- Interfacial force is determined by the chemical properties of the two phases which are in contact, especially the chemical properties of the boundary layer of the two phases. It is also affected by the temperature and pressure of the two phases.

Since σ is sensitive primarily to the chemical properties of boundary layers, any chemical constituent which tends to accumulate preferentially in the boundary layer, may have a large effect even though the average concentration in the phase as a whole may be very small. Such agents are said to be surface active and are called surfactants. Surface active agents which reduce σ for water, for example, occur frequently as contaminants in soils, rocks and in laboratories. Most often they consist of some type of organic molecules.

The reasons why many organic molecules are surface active can be understood by considering the following principles:

(1) The tendency of the system as a whole is to reach a condition of minimum energy. Other things being equal this means a minimum surface energy.
(2) In a solution, the ordinary tendency is for molecules or ions to mix homogeneously, but the effect of interfacial energy is to oppose this tendency.
(3) With a mixture of constituents in a solution, σ is determined primarily by the molecules actually in the surface.
(4) Molecules or ions in the surface layer are under the influence of interfacial force. Therefore, they possess more energy than interior molecules.
(5) It requires energy to move molecules from interior to surface.
(6) With a mixture of constituents, the minimum surface energy is obtained when those constituents which result in a lower surface tension have a higher concentration in the surface.

Based upon the above considerations, J. Willard Gibbs (1876) derived an equation relating change in σ with concentration of particular chemical agents. He showed that

$$\frac{d\sigma}{dc} = - RT \frac{\Delta c}{c} \tag{1.7}$$

in which

Δc = excess concentration of agent in surface layer,
c = concentration of agent in bulk of liquid phase,
R = gas constant,
T = absolute temperature.

Equation 1.7 reflects the fact that an agent which tends to concentrate in the surface, so that Δc is positive, causes σ to decrease with respect to c. Therefore, an agent that tends to reduce σ will accumulate preferentially in the surface, because by accumulating largely in the surface, σ will be reduced in conformance with the principle of minimum energy at equilibrium.

An agent that tends to increase σ will have a smaller concentration in the surface layer than in the remainder of the fluid in conformance with the same principle, and Δc will be negative. Since only those constituents in the surface layer have a significant effect on σ, the concentration c must be larger to produce a substantial effect in the latter case. Consequently, agents that tend to increase σ cannot be called surfactants. Most inorganic salts are in a class of chemicals the effect of which is to produce a small increase in σ of water solutions in contact with air.

Surfactants usually consist of relatively large molecules, often organic. Large molecules which are polar and which tend to orient themselves to exclude other molecules in the surface are particularly effective in this respect. Many such materials consist of organic salts of inorganic ions. An example is sodium sulfonate. The Na^+ ion tends to dissolve in a water solution leaving a negatively charged sulfonate group (anionic) oriented in the surface with the negative end pointed toward the interior of the solution. Other surfactants are cationic, that is, the surface active part possesses a positive charge, and some are nonionic.

Almost any organic contaminant will reduce σ of pure water. In fact, σ of pure water is very unstable. This is because σ of water is greater than that for most common liquids. Consequently, the value of σ for water in soils and porous rocks is less than that specified in handbooks for water at a particular temperature. Water in naturally occurring earth materials often has a value of $\sigma \cos \alpha$ of about 60 dynes/cm, whereas σ for pure water in contact with air at 20°C is about 72 dynes/cm. In contrast, most oils have a smaller σ so that they are not so easily affected by common contaminants. For example, an oil commonly used for laboratory flow studies which bears the trade name of Soltrol* has a σ of approximately 23 dynes/cm at laboratory temperatures. It is used because its surface tension and wetting properties are much more stable than those of water.

*Manufactured by Phillips Petroleum Inc., Bartlesville, Oklahoma.

Temperature also affects surface tension. The variation of σ of water in contact with air at room temperatures is small, being about 75 dynes/cm at 5°C and 72 dynes/cm at 25°C. At higher temperatures, the change of σ in respect to temperature increases, and at 100°C, the σ of water-air is about 50 dynes/cm. Some representative examples of other surface tensions in dynes/cm at 20°C include:

mercury-air	≅ 470,
mercury-water	≅ 375,
water-octane	≅ 51,
ethyl alcohol-air	≅ 21.5,
water-octyl alcohol	≅ 8.5.

1.5.3 Wettability - Interfacial force also exists at solid-fluid interfaces, although it is extremely hard to measure. The interfacial force at solid surfaces is important, because it controls which fluid is the wetting or the non-wetting fluid. It also determines the contact angle of the interface with the solid as Equation 1.3 implies. An associated property of solid surfaces is the energy required to remove adsorbed constituents of various fluids, that is, the "strength" of adhesion.

Such properties are one aspect of what is called wettability. Another aspect of wettability is the speed with which fluids will spread over solid surfaces. The speed of spreading is affected by the surface tension and also by viscosity. Spreading speed is increased by lower surface tensions and lower viscosities.

It may sometimes happen that when two liquids such as water and oil are introduced at the same time into a porous medium, the surface will first be wet by the oil and later the oil will be replaced in the adsorbed layer by water. The initial wetting by oil is due to its smaller surface tension in contact with air which results in oil spreading faster than water. The reason for oil being eventually replaced by water is that water is more strongly adsorbed.

The chemical condition of solid surfaces, particularly in respect to adsorbed organic contaminants, is important in determining the strength of adhesion of particular fluids and, to some extent, the speed of spreading. Petroleum scientists frequently classify porous rocks as being either water-wet or oil-wet, depending upon which of these liquids preferentially becomes adsorbed on the solid. Given sufficient time, however, water will replace most oil constituents on the solid surfaces of nearly all porous rocks. The condition of being oil-wet is usually a temporary situation existing in laboratory samples after they have been dried and when they have a residue of adsorbed organic contaminants remaining. Most of these contaminants can be removed by heating the samples to the ignition temperature of the contaminant.

Some adsorbed organic molecules, however, are very difficult to remove except by burning. Bradford sandstone in Pennsylvania, famous for the production of paraffin based crudes, from which lubricating oils are refined, is a well-known example of a rock considered to be oil-wet in its natural state. In some cases, porous materials may be deliberately made water-repellant by the adsorption of agents such as silicon on their surfaces.

Agricultural, lawn and forest soils sometimes become temporarily water-repellant after desiccation because of the presence of organic substances, especially bitumens produced by the metabolic processes of certain microorganisms. Prolonged exposure to water will often restore the normal hydrophilic character of the soil. Exposure of the soil surface to heat, for example from a torch, will remove the bitumens faster.

REFERENCES

Adam, N. K. (1968), The Physics and Chemistry of Surfaces. Dover Publications, New York, N.Y., "Unabridged and corrected republication of the third (1941) edition." pp. 1-16.

Bower, C. A. and Goertzen, J. O. (1959), Surface area of soils and clays by an equilibrium ethylene glycol method. Soil Science, Vol. 87, No. 5, May, pp. 289-292.

Collins, R. E. (1961), Flow of Fluids Through Porous Materials. Reinhold Publishing Corporation, New York, 270 p.

Donaldson, E. C., Kendall, R. F. and Baker, B. A. (1975). Surface-area measurement of geologic materials. Society of Petroleum Engineers Journal, April, pp. 111-116.

Encyclopaedia Britannica Inc. (1964 ed.), Surface tension, Vol. 21, pp. 595-604.

Gibbs, J. W. (1876), On the equilibrium of heterogeneous substances. Transactions of the Connecticut Academy of Sciences, Vol. 3, pp. 228-391.

PROBLEMS AND STUDY QUESTIONS

1. By knowing the specific gravity of sand particles, usually about 2.65 and the total volume in which a given weight of this sand is contained, describe how the average porosity of the unconsolidated sample could be estimated.

2. Could a similar method of estimating average porosity (as defined in this text) be used to estimate the porosity of a volcanic rock containing many void spaces formed from gas bubbles at the time of lava cooling? Explain.

3. Explain why a geological formation consisting of cavernous limestone rock might not behave as a porous medium as defined in this text.

4. Agricultural soils usually have much greater porosities than deposits of sand on beaches. Explain.

5. Describe the effect (of adding a relatively small amount of montmorillonite to a sample of beach sand) on the average pore size. Explain.

6. Explain why it is appropriate to apply the term "average pore-size" for the ratio ϕ/s.

7. Would you expect the addition of montmorillonite to beach sand to have any significant effect upon the pore-size distribution? Explain.

8. Would you expect average porosity of a consolidated sandstone rock (containing a small amount of clay) to be affected to a greater or lesser extent (by clay dispersion) than a pore-size distribution index? Explain.

9. Explain why a "fluid particle" in a gas system is necessarily larger than a "fluid particle" in a water system.

10. Derive, by means of the "balance of work" principle, an equation for the excess pressure inside a spherical air bubble surrounded by liquid.

11. Determine the pressure intensity within a soap bubble 0.01 cm in diameter. Assume σ of the soap solution, at air interfaces, is 40 dynes/cm.

12. Explain why σ for the soap solution is 40 dynes/cm and not about 70 dynes/cm as would be the case for pure water.

13. Contrast the trends of pressure variation during the process of: a. blowing up a toy ballon, and b. blowing a soap bubble. Explain.

14. In what physical way does an oil-gas two-phase system differ from a brine-fresh water system encountered in a salt water intrusion into a coastal aquifer? Explain.

15. Explain how the relative strength of adhesion to a solid surface of two fluids (mixed in a porous medium) will determine which fluid phase will be at the higher pressure.

16. Why is it necessary to consider a system of two immiscible fluids as two separate continua in the application of fluid mechanics?

17. A porous medium sample consisting of a gallon full of beach sand may have either a slightly larger or a slightly smaller porosity than a gallon can full of uniform marbles. Describe what factors might make the porosity of the sand greater and what factors might make it less than that of the marbles.

18. Agronomists often refer to water in a soil profile within the root zone of plants as being under "tension". Is this use of the term "tension" analogous in any way to the use of this term in connection with the state of stress in a steel bar? Explain.

19. Would you consider the term "suction" more (or less) appropriate than "tension" to designate capillary pressure? Explain.

20. Would you expect a water-air interface or a water-oil interface to have a smaller angle of contact? Explain.

21. Explain the contrast in orientation of interfaces illustrated in Figure 1-1.

22. Considering the principle of minimum energy for equilibrium, would you expect (with a very small concentration of amyl alcohol) to find all of the alcohol molecules in the surface layer? Explain the principle involved in terms of thermodynamic concepts.

23. Give a possible reason why σ for mercury-air is greater than that for mercury-water and why both are greater than that for water-air.

24. It sometimes happens that in late summer, lawns develop dry spots that seem hard to wet. By poking a few holes through the turf in these spots, the soil can be made to absorb water easily. Explain.

25. Consider a cylindrical sample of porous sandstone (granular) having a diameter of 2.54 cm and a length of 6 cm. The sand grains have a specific gravity of 2.65. Before drying, the sample weight is 60 gm. After drying, the weight is 53 gm. Assume that the liquid contained in the sample before drying was entirely water. Estimate ϕ, S, θ, and W before drying.

26. In reference to the sample described in problem 25, describe a method of checking the estimated porosity using an independent procedure.

27. Given that the solid grains contained in the sample of porous sandstone of problems 25 and 26 have a specific surface measured by the ethylene glycol procedure to be 10 m^2/g, estimate the average specific surface of the bulk sample in cm^2/cm^3.

CHAPTER II

HETEROGENEOUS FLUIDS IN STATIC SYSTEMS

2.1 MECHANICAL EQUILIBRIUM

A fluid system is said to be in equilibrium when there is no net transfer of matter within the system and no flow of heat.

The conditions for equilibrium may be studied from the point of view of thermodynamics. A thermodynamic description of a fluid system begins with a definition of a system which in this case is a definite quantity of fluid that is in interchange with the surroundings only by flow of heat or by doing work. In the case under consideration, the porous solid is the surroundings.

According to Zemansky (1943) there are three conditions that must be satisfied for any system to be in equilibrium:

(1) There must be no unbalanced driving force on any element of the system and none between the system and its surroundings. This is a condition for mechanical equilibrium.
(2) The fluid system must not undergo a spontaneous change of internal structure such as a transfer of matter from one phase to another. A system that meets this requirement is in chemical equilibrium.
(3) All parts of the system must be at the same temperature and this temperature must be the same as that of its surroundings. Such a system is in thermal equilibrium.

When conditions for all three types of equilibrium are satisfied, the system is said to be in a state of thermodynamic equilibrium. States of thermodynamic equilibrium can be described in terms of macroscopic coordinates that do not involve time, that is, thermodynamic coordinates. These coordinates must refer to the system as a whole and not to its parts.

If an attempt is made to describe a fluid system in terms of thermodynamic coordinates, the system must be relatively small, otherwise the description will be inadequate. Furthermore, when equilibrium does not exist, coordinates referring to a system as a whole may not be possible unless the system selected is very small. This is because gradients of pressure and temperature exist.

The approach used in fluid mechanics is to choose a system consisting of either a very small control volume which remains fixed in space or a small element of fluid which may move in space. The latter approach is employed here.

In order to limit the scope of phenomena considered to an entity of reasonable size, only problems relating to mechanical equilibrium are considered in detail. The analyses presented assume that chemical and thermal equilibrium exists.

A fluid is said to be static when all of its elements are in mechanical equilibrium. In such a state, none of its elements move with respect to a coordinate system fixed to the solid boundaries, because the driving forces on each element are balanced.

A fluid particle as defined in Section 1.5.1 is selected as a reference element (or free body) for the analysis of forces acting in a static fluid. A fluid particle is considered which is entirely within a single fluid phase. Consequently, there is no ambiguity in respect to density and other properties assigned to the particle. It is regarded as being a part of a single continuum constituting one fluid phase which may be mixed with another phase (or other phases) occupying the pore space.

In some cases, the non-wetting fluid may be divided into separate parts completely surrounded by the wetting phase. When a phase is discontinuous in this sense (not interconnected) it may be regarded as a part of the porous matrix, analogous to the solid grains, in respect to an analysis of the wetting phase. When two or more interconnected phases are present, each is analyzed separately as a single continuum.

2.1.1 Forces on static fluid particles - Forces acting on a reference particle are classified as driving or resisting forces according to whether they tend to produce motion or are a consequence of motion. In the case of static fluids, only driving forces act on fluid particles. Furthermore, their resultant must be zero.

Driving forces, in turn, can be classified according to whether they are proportional to the mass of the particle or to the surface area of the particle, the former being called body forces and the latter surface forces.

The most conspicuous of the body forces acting on fluid elements is gravity. For the present analysis, it is assumed that this is the only body force. Force per unit volume due to gravity is $\rho \underline{g}$ where ρ is fluid density and $\underline{g}$ is a vector representing force per unit mass due to gravity, the line under the symbol indicating a vector quantity. The resultant of $\underline{g}$ is directed vertically downward through the center of mass of fluid particles. Although $\underline{g}$ varies inversely as the square of the distance between the fluid particle and the center of the Earth's mass, it is sufficient to regard g as a constant for applications considered here. A value for g of 980 dynes/gram is close enough for most purposes.

The force acting on the surface of fluid particles, in general, can have tangential as well as normal components. Tangential components, however, exist only if there is relative motion of the center of one fluid element in respect to another. This does not happen in a static fluid.

The normal component of surface forces may be a result of two factors. The first factor called pressure or hydrostatic pressure is that normal component which can be related in an equation of state to the density and temperature of the fluid. This portion of the normal surface force is conservative; that is, it has a fixed value for a given fluid at a particular temperature and density. In the case of static fluids, pressure is the only surface force and it acts equally in all directions. From a microscopic point of view it can be regarded as resulting from the rate of momentum transfer (by fluid molecules) normal to any differential area at a point in the fluid. In static fluids pressure is invariant in respect to orientation of the differential area.

Pressure, therefore, is considered to be a scalar quantity and to act at a "point". It must be defined, however, in respect to a volume element (the size of a fluid particle) containing the point. The definition is given by

$$p \equiv \lim \frac{1}{A} \int_A \underline{\sigma}_c \cdot \underline{dA} \qquad 2.1$$

A → A of fluid particle

in which A is the magnitude of the surface area of a fluid volume and $\underline{\sigma}_c$ is the conservative surface stress, positive outward. Note that $\underline{\sigma}_c$ is a vector quantity whereas p is a scalar.

Another part of the normal surface force results from viscous resistance to expansion or compression of fluid particles. It does not exist in static fluids or in flowing fluids which are not undergoing expansion or contraction, that is, in which the divergence is zero. The portion of normal surface stress which is not conservative, and is associated only with divergence, is not related to density at a particular temperature. In the case of flowing fluids, even those undergoing negligible divergence, $\underline{\sigma}_c$ is not, in general, invariant in respect to direction. In such cases Equation 2.1 is still a valid definition of p and can be shown to be equivalent to

$$p = \frac{1}{3}(\sigma_i + \sigma_j + \sigma_k) \qquad 2.2$$

in which σ_i, σ_j and σ_k are normal surface-stress components in three orthogonal directions. In this case, also, p is regarded as a scalar quantity, although it has the dimensions of force per unit area.

For the analysis of static fluids, therefore, the only forces considered to act on fluid particles are gravity and the resultant of the pressure force. Because pressure is a scalar, the resultant force on a fluid particle due to pressure is entirely due to the variation of pressure in space. Specifically, the component of force per unit volume due to pressure (in a particular direction i) is given by

$$-\frac{\partial p}{\partial x_i} ,$$

x_i being a coordinate length in the direction i, and the minus sign indicating that the force is positive in the direction i if p decreases in that direction.

A balance of force components in the direction i is given, therefore, by

$$-\frac{\partial p}{\partial x_i} + \rho g_i = 0 , \qquad 2.3$$

in which g_i is the component of gravity in the direction i. Note that the force components in Equation 2.3 are expressed as force/volume.

2.1.2 Forces on fluid in a control volume - The preceding analysis is in respect to a reference element consisting of a fluid particle. Some authors prefer to consider a force balance on a fluid volume contained in an element V_m such as is used for the definition of porosity at a point in Section 1.4.1. To show that the latter approach also leads to Equation 2.3, a greatly simplified model of an element V_m is considered, which is illustrated in Figure 2-1.

The reference volume V_m consists of both solid and fluid space, but only forces acting on the fluid contained in V_m are considered. In this case, a single fluid is assumed to occupy all of the pore space.

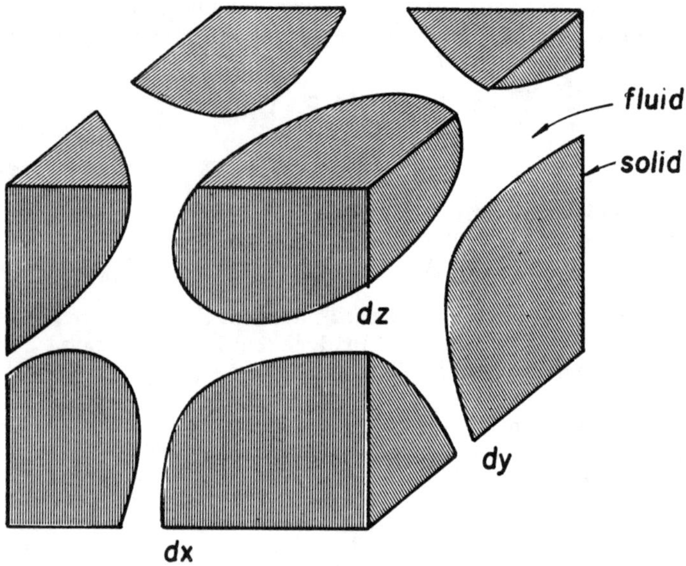

Figure 2-1. Cross-section of model for reference volume V_m.

Force due to gravity is $\rho \underline{g} \phi V_m$, or if only the component in a particular direction i is considered, $\rho g_i \phi V_m$.

The force resulting from unbalanced fluid pressure on the faces of V_m can be visualized by considering that which exists on two opposite faces, say top and bottom. For this purpose, a situation is considered in which porosity may be varying in space so that the fluid area exposed at the upper surface of V_m is ϕ_u dx dy, and that at the lower face is ϕ_ℓ dx dy. The total force due to fluid pressure on the upper face is $p_u \phi_u$ dx dy and that on the lower face is $p_\ell \phi_\ell$ dx dy. The net component of force in the positive z direction is

$$(p_\ell \phi_\ell - p_u \phi_u)\, dx\, dy \quad .$$

Insofar as it is permissible to regard p and ϕ as point concepts, the component in the z direction due to fluid pressure on the reference element is

$$- \frac{\partial(p\phi)}{\partial z}\, dz\, dx\, dy \quad ,$$

which expressed as force/volume is

$$- \frac{\partial(p\phi)}{\partial z} \quad .$$

Generalizing, in three dimensions, the force due to fluid pressure in any direction i is given by

$$- \frac{\partial (p\phi)}{\partial x_i} .$$

However, since the reference element is the fluid contained in the volume V_m it must be remembered that the solid surface (as well as fluid surface) exerts a force on the fluid within V_m. If porosity of the medium is uniform so that the fraction of solid on each face of V_m is the same, the resultant of the solid surface force on the fluid element is zero. However, if ϕ is not uniform, the resultant of this force is not zero.

This situation can be visualized by expanding the expression for the force/volume exerted by the fluid, that is,

$$- \frac{\partial (p\phi)}{\partial x_i} = - \phi \frac{\partial p}{\partial x_i} - p \frac{\partial \phi}{\partial x_i} .$$

Clearly, the first term on the right represents a force exerted by fluid on only the fluid in the reference element. The second term on the right represents a force exerted by fluid on solid surfaces of the reference element. However, the force of fluid on solid surfaces is exactly counterbalanced by an opposite force of solid surfaces on the reference fluid element. Consequently, the net surface force/volume is given by

$$- \phi \frac{\partial p}{\partial x_i} - p \frac{\partial \phi}{\partial x_i} + p \frac{\partial \phi}{\partial x_i} = - \phi \frac{\partial p}{\partial x_i} .$$

Summing the pressure force and the gravitational force and canceling ϕ results again in Equation 2.3.

2.1.3 Adsorptive forces - The foregoing analyses of forces on static fluid particles assume that the only body force of significance acting on the particles is gravity. It has been pointed out by Edlefsen and Anderson (1943) that this is not necessarily a valid assumption, especially in reference to a wetting phase at low saturations.

Forces at fluid-solid interfaces which produce adsorption and wetting may extend outward from the solid surface for a few hundred molecular layers. Although such forces are known to be very strong, they are also very short ranged. It has been estimated that they vary inversely with approximately the 5th power of distance from the solid surface. If a particle is more than say 100 molecular diameters from a solid surface, the adsorptive force $\underline{a}$ can probably be neglected. But, the potential energy due to $\underline{a}$ cannot necessarily be regarded as small.

The magnitude of $\underline{a}$, which may be electrostatic in origin or due to a variety of other possible factors, depends upon the species of molecules in a fluid particle. It is not entirely clear that such forces can always be treated as a body force, but following the precedent of Edlefsen and Anderson (1943) this assumption is made in respect to a fluid particle illustrated in Figure 2-2.

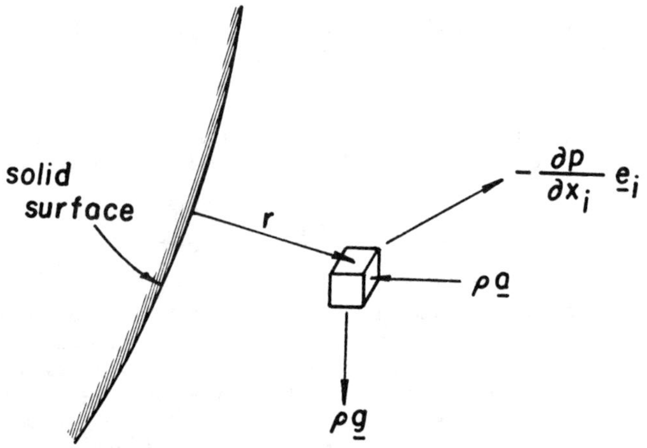

Figure 2-2. A fluid particle in an adsorptive force field.

In Figure 2-2 a fluid particle is shown which is a distance r from a solid surface. The forces, expressed as force/volume, acting on the particle are indicated. If the potential energy per volume in respect to these force fields are summed and the sum is designated as p*, the result is

$$p^* = p + \rho \int_{\underline{d}}^{\underline{s}} -\,\underline{g} \cdot d\underline{s} + \rho \int_{\underline{d}}^{\underline{s}} -\,\underline{a} \cdot d\underline{s}$$

where $\underline{s}$ is a positional vector, $\underline{d}$ represents $\underline{s}$ at a datum point, and ρ is the fluid density treated here as a constant. For practical purposes, the first integral on the right can be represented by ρgh. If the particle does not move within say 100 molecular layers of the solid surface, the second integral is essentially zero in respect to the datum point selected. In this case, the potential energy in respect to the three force fields is given by

$$p^* = p + \rho g h$$

which is the familiar piezometric pressure.

If the particle moves into a range where $\underline{a}$ is large, the second integral may take on a large negative value. If this integral is neglected, an error may be made in determining the value of p at points in the pore space. For example, it might be concluded that water is under a large negative pressure (tensile stress), whereas it is only the last integral that is negative. A correct analysis, if that would be possible, might show that the water is actually under a substantial positive pressure.

The latter conclusion seems to be justified experimentally because it is known that water in porous media at low saturations has a lowered freezing point which is characteristic of water under positive pressure. On the other hand, the water has a reduced vapor pressure which indicates the opposite. However, the latter situation can be explained by the probability that adsorptive force fields have a similar effect on the escaping tendency of water molecules as a reduced water pressure.

The reader should observe that the foregoing analysis is by no means rigorous since the concept of a fluid particle as previously defined actually does not apply in respect to fluid within a few hundred molecular layers from a solid surface. Consequently, the analysis must be viewed only as an artifice to aid in the visualization of the qualitative effect of adsorptive forces.

Furthermore, it is not possible to measure the separate potential energy resulting from adsorptive forces. Instruments for measuring fluid pressures in porous media actually measure the sum of pressure and the potential energy due to adsorptive forces. Unless one is concerned about problems dealing with freezing point depression, or something similar, this is of no consequence. If the fluid under consideration is under the influence of adsorptive force fields, it is the sum of p and the potential energy (due to adsorptive force) which controls the mechanical equilibrium of the fluid. In such cases the measured p represents this sum and may be regarded as an apparent pressure. In the remainder of this text, p is used to designate the measured pressure whether it happens to be the actual pressure or only an apparent pressure.

2.2 DISTRIBUTION OF PRESSURE IN A STATIC TWO-PHASE SYSTEM

Since the resultant of $\underline{g}$ is vertically downward, the force balance can be written as

$$- \frac{dp}{dz} - \rho g = 0$$

in which g is a scalar quantity representing the magnitude of $\underline{g}$.

The solution of the preceding equation is

$$p - p_o = - \rho g h \qquad 2.4$$

in which h is the elevation above a datum where p is p_o. Equation 2.4 gives the pressure distribution in a static fluid system. When applied to heterogeneous fluids in a porous medium, however, it must be applied simultaneously to each of two or more fluids. For example, in the case of a two-phase system

$$p_w - p_{w_o} = - \rho_w g h$$

and

$$p_{nw} - p_{nw_o} = - \rho_{nw} g h$$

where the subscripts w and nw refer to the wetting and non-wetting phases respectively. Subtracting the first from the second of these equations gives

$$p_c = (\rho_w - \rho_{nw}) g h + p_{c_o} \qquad 2.5$$

in which p_c is the capillary pressure at an elevation h, and p_{c_o} is the value of p_c at the datum from which h is measured. Provided each phase is interconnected over the interval h, Equation 2.5 gives the value of p_c at all points of contact between phases.

A variation of Equation 2.5 is often used in reference to water in soil profiles under static conditions; that is,

$$p_c = \rho_w g h \ , \qquad 2.6$$

the density of air being taken as zero, and h being measured above a datum where p_w is at zero gauge pressure. The locus of points at which water is at atmospheric pressure is called a <u>water table</u>. In a petroleum reservoir, none of the fluids are at atmospheric pressure so Equation 2.6 is not applicable. Equation 2.5, however, can be used for any two-phase fluid system in a static condition.

2.3 DEPENDENCE OF SATURATION UPON CAPILLARY PRESSURE

Combining Equation 1.6 with Equation 2.5 results in

$$\sigma\left(\frac{1}{r_1} + \frac{1}{r_2}\right) = \Delta\rho\ gh + p_{c_o} \quad ,$$

indicating that in a static two-phase system, the mean curvature at points on interfaces changes with elevation. Specifically, the radii of curvature become smaller at higher elevations. It is the change in curvature which permits the greater pressure discontinuity across interfaces at higher elevations.

The changes in curvature and p_c are accompanied by a change in fluid distribution. Saturation of the wetting phase becomes smaller and that of the non-wetting phase becomes greater as elevation increases. A functional relationship between S and p_c can be visualized by considering a model of a cross-section of an element of pore space as illustrated in Figure 2-3. In this case, the pore space contains a mixture of water and air.

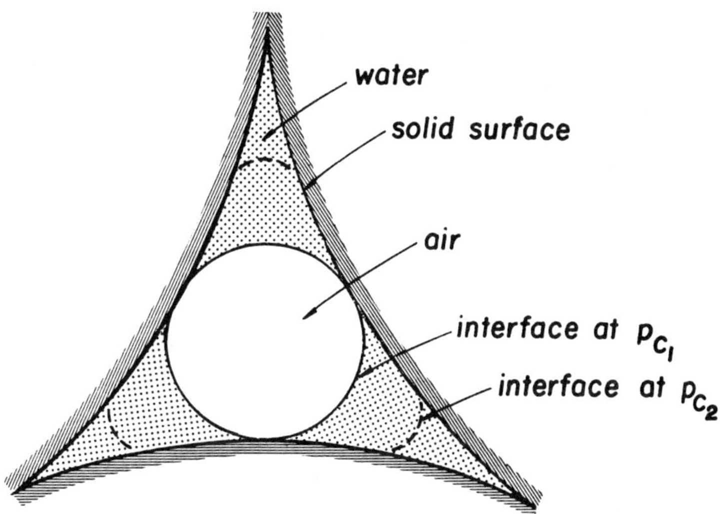

Figure 2-3. Model of pore cross-section with varying saturation of water.

As p_c is increased from p_{c_1} to p_{c_2}, either by increasing the air pressure or decreasing the water pressure, a volume of water is removed from the pore space. In the process, the interfaces retreat to portions of the pore space having smaller dimensions so that the radii of curvature are smaller. In other words,

$$S = f(p_c) \quad .$$

However, if the pore space initially is fully occupied with water, a finite value of p_c (designated at p_e) must be exceeded before air can intrude into this element of the pore volume. The value of p_e, called air-entry pressure, depends upon dimensions of the largest "opening" into the particular element of pore volume considered. If the pore volume considered contains some portion of pore space with dimensions larger than the largest "opening," that portion of the pore space will immediately desaturate. Consequently, the desaturation of pore space with increasing p_c, in its initial stages, does not occur smoothly but proceeds in "jumps" which sometimes can be observed experimentally [Corey and Brooks (1975)].

The "jumps" are observed by the sudden entry of air into portions of the medium where none existed previously, and also by small but abrupt fluctuations in water pressure. The latter phenomenon is caused by the sudden discharge of water from certain portions of the pore space, resulting in a slight and temporary increase in pressure of the surrounding water. The fluctuations have a greater amplitude in media having a wider range of pore sizes. After practically all of the pore space contains some air, further increases in p_c may proceed without noticeable fluctuations.

If a laboratory sample of porous medium is caused to desaturate by increments of p_c, and allowed to arrive at a static state with each increment, the values of S determined for corresponding values of p_c will provide a curve as shown by the solid line in Figure 2-4.

In the petroleum literature, it is customary to plot p_c as a function of S as has been done in Figure 2-4 [Richardson (1961)]. Such curves are called capillary pressure-saturation curves [Collins (1961)]. Soil scientists usually plot θ as a function of ψ, but the relationship is equivalent, it being entirely arbitrary as to whether S or p_c is regarded as the dependent variable. Soil scientists refer to such curves by a variety of names including water-retention and water-characteristic curves.

2.3.1 Hysteresis in the capillary pressure - Saturation Function

- The relationship determined for $p_c(S)$ is not unique, even for constant values of σ and α, but depends on the pressure (or saturation) history. In other words, it is subject to hysteresis. A curve such as the solid line in Figure 2-4 is obtained starting with a wetting phase saturation of 1.0. This may be called a desaturation curve in the petroleum literature or a water-release, drainage or drying curve by soil scientists.

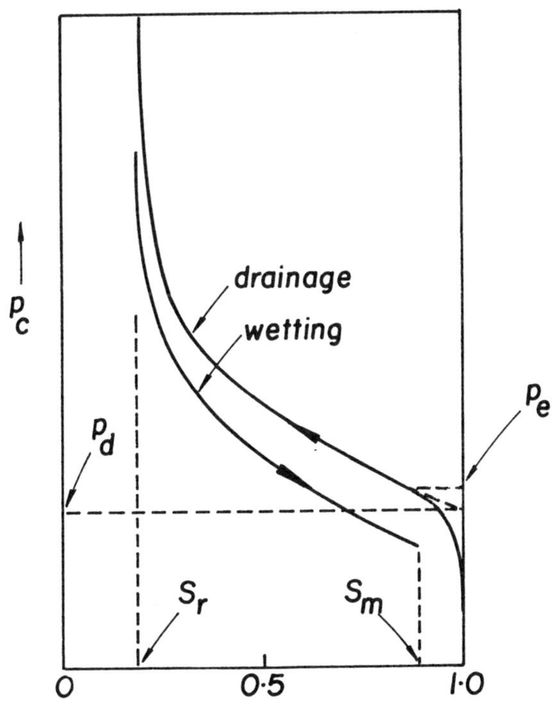

Figure 2-4. Capillary pressure as a function of saturation.

Another curve is obtained by starting with a sample containing only the non-wetting phase and allowing it to imbibe the wetting phase from a source which undergoes successive increases in pressure relative to that of the non-wetting phase. It is sometimes referred to in petroleum literature as an imbibition curve and in the soils literature as a wetting curve.

The two curves shown belong to an infinite family of curves that might be obtained by starting at any particular S_w

and either increasing or decreasing S_w. All such curves, presumedly, would be between the two illustrated. Hysteresis has been studied intensively by a substantial number of investigators including Topp (1969).

One factor which undoubtedly is involved in hysteresis is the matter of wettability changes depending upon which phase is in contact with the solid first. Another factor, frequently cited, is illustrated in Figure 2-5 which shows two identical capillary tubes with irregular cross-sections. On the left is illustrated a tube first filled and then allowed to drain into the liquid reservoir, whereas the tube on the right was initially empty and then allowed to imbibe liquid from the reservoir. The capillary pressure at the liquid-air interfaces in the tubes is given by Equation 1.5 and the height of rise above the reservoir by Equation 2.6. Since the water stands about

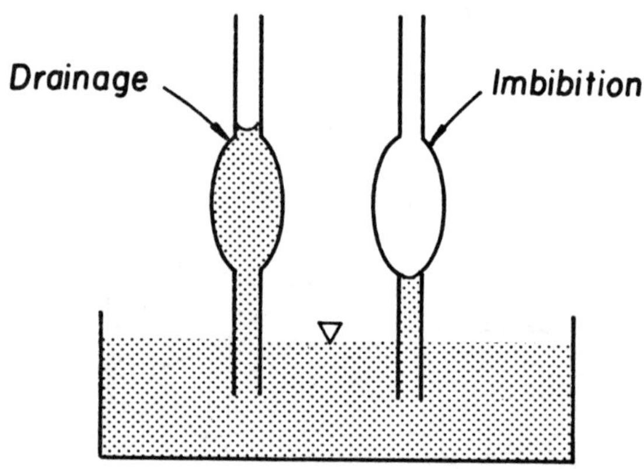

Figure 2-5. Capillary hysteresis.

twice as high in the tube initially full, the p_c for this tube must be about twice as large. The drained tube, however, contains much more liquid, because the interface in the tube on the right cannot advance beyond the enlarged cross-section.

The situation in ordinary porous media is, of course, much more complex than indicated by the simple model, but it is supposed that an analogous mechanism exists in soils, and porous media always contain more water on a drainage than on a wetting cycle.

Note that values of $p_c(S)$ are not shown in Figure 2-4 for values of $S > S_m$ on the wetting cycle. The reason is that at the critical saturation S_m, the non-wetting phase becomes entrapped. It is no longer interconnected and cannot be replaced simply by decreasing p_c. Since the non-wetting phase is not interconnected, it does not have a unique or a measurable pressure so that p_c cannot be determined. Soil scientists sometimes present data for this range of S, but in that case the data represent the negative pressure of water relative to atmospheric pressure, rather than relative to the entrapped air phase.

2.3.2 Entry pressure - As Figure 2-4 indicates, on a drainage cycle, S is very close to 1 over a finite range of p_c. When some critical value of p_c is exceeded, S decreases rapidly with increasing p_c. It is supposed [White et al. (1972)] that desaturation occurring at p_c smaller than the critical p_c takes place in pore space exposed at the sample boundary. The rationale for this view is that the non-wetting phase cannot reach the interior of a sample until an interconnected network of channels have been desaturated.

It was shown by White et al. that the inflection point on the curve, where dS/dp_c is a maximum, corresponds to the highest saturation at which the non-wetting phase is interconnected, that is, at which non-wetting phase permeability exists.

There is some ambiguity in the literature in regard to what point represents the critical value of p_c on the curve of $p_c(S)$. Petroleum scientists define a critical p_c, called displacement pressure p_d, as being the p_c at which first desaturation on a drainage cycle occurs. However, since some desaturation takes place in laboratory samples at all values of $p_c > 0$, it is difficult to decide at what p_c a significant desaturation takes place. Usually, petroleum scientists appear to have arrived at what they call p_d by extrapolating the curve to the ordinate where S is 1, neglecting that part of the measured data which are for values of S near 1. In the remainder of this text, this is the interpretation assigned to p_d.

It is common practice in the ceramics industry to classify porous ceramics according to the air pressure needed to force air through an initially water-saturated sample. They call this the bubbling pressure p_b. Presumedly, this corresponds to p_c at the inflection point as shown in Figure 2-4. Soil scientists define a similar parameter as the air-entry pressure p_e. Since both bubbling pressure and air-entry pressure imply that the non-wetting phase under consideration is air, and since the term displacement pressure has not been consistently employed, a better term for general application might be simply entry pressure.

In most cases, the value of S at p_e is in the range 0.8 to 0.9, but p_e may vary over an enormous range, from practically zero to many atmospheres.

2.3.3 Residual saturation and effective porosity - Another parameter of considerable significance is the value of S at which p_c increases very rapidly with negligible decrease in S. This is called the residual or irreducible saturation in the petroleum literature. A somewhat analogous concept in respect to a water saturation found in an otherwise oil-saturated rock is called connate water. Soil scientists sometimes refer to a related parameter θ_{min} as the minimum water content. The term residual saturation S_r is used in this text because the words "irreducible" or "minimum" seem to imply a physical meaning which is not intended.

For example, it clearly is possible to remove practically all of the wetting fluid from a porous sample by evaporation. Furthermore, plants may remove water from soils at a saturation substantially less than S_r. In many cases, even if the process of removal is restricted to liquid flow, there is no well-defined minimum saturation. In spite of its ambiguity, however, the concept of a residual saturation has great practical utility. Corey (1954) and Brooks and Corey (1966) have presented methods of finding a value of S_r more or less objectively by extrapolation processes using data such as are shown in Figure 2-4. These are discussed in Section 2.4. Sometimes S_r is determined as the value of S at some arbitrarily large p_c.

The physical interpretation of S_r depends to some extent upon how it is determined. Some authors have claimed that S_r represents a saturation such that wetting phase films over the solid surface are no longer interconnected, and that the wetting phase exists only in pendular rings about points of contact of grains, or other isolated spots. This view does not seem to be tenable in view of the observed fact that the wetting phase never becomes completely immobile and the curve of $p_c(S)$ never becomes exactly vertical.

When S_r is found by extrapolation, the result seems to be associated with a volume of pore space characterized by pore sizes substantially smaller than that of the bulk of the pore space. Many media, porous sandstones being a typical example, frequently yield S_r values that are correlated with the amount of clay in the sample. The residual saturation can be reduced to a very low value, often less than 0.05, by removing the clay with an ultrasonic cleaning technique. Evidently, pore sizes associated with the clay are orders of magnitude smaller than those associated with the remainder of the pore space enclosed by the sand grains.

Furthermore, when S_r is found by extrapolation, it often happens that S_r is zero, or even slightly negative. A zero value of S_r occurs commonly in fine-grained highly structured soils with a very wide pore-size distribution, even though they usually have a relatively large clay content. Evidently the extrapolated value of S_r may be related to a discontinuity in pore sizes such that sizes (associated with $S < S_r$) are distinctly smaller than those corresponding to $S > S_r$. In media with a pore size ranging smoothly from zero to some finite value, S_r appears to be zero.

In summary, S_r, when determined as the S at an arbitrarily high p_c, is ambiguous because it depends upon the value of p_c selected. When S_r is determined by an extrapolated procedure, its physical meaning is uncertain and it probably should be interpreted as a parameter useful for curve-fitting purposes as explained in Section 2.4.1.

Because the pore space containing the wetting phase as $S < S_r$ contributes <u>relatively</u> very little to convective flow processes, it is convenient (for some purposes) to define an <u>effective saturation</u> S_e given by

$$S_e \equiv \frac{S-S_r}{1-S_r} \quad . \qquad 2.7$$

Obviously, S_e is significant mostly in reference to flow problems, but it is also useful in empirical representations of $p_c(S)$, as is explained in Section 2.4.1. Similarly an <u>effective</u> or <u>drainable porosity</u> is defined as

$$\phi_e = (1-S_r)\,\phi \quad . \qquad 2.8$$

2.3.4 Field capacity - A concept somewhat related to residual saturation is called field capacity by agronomists and soil scientists (Miller and Klute, 1967). The term is used for two somewhat different concepts. The first is often referred to simply as "field capacity" F.C. and means the water content W in a soil profile after "downward drainage has become very slow" following a thorough wetting. Defined in this way, F.C. may be affected by the position of a water table, if one is present, and by the structure of the entire soil profile as well as the properties of the soil in a particular stratum. Sometimes F.C. is taken as the value of W after an arbitrary period of time following a thorough wetting. A method of measuring F.C. in the field has been described by Peters (1965).

A second usage of the term field capacity may sometimes be called laboratory field capacity or the normal moisture capacity (Shaw, 1927). In this case, it is determined on a column of soil in a laboratory. A substantial amount of water is added to the top of a column of dry soil, but not enough to wet the soil to the bottom of the column. The entire column is then allowed to reach a pseudo static equilibrium while being protected from evaporation. Values of W are then determined for samples of the soil at elevations sufficiently far from the dry soil to be unaffected by the water content gradient in that region.

Values of laboratory field capacity, when converted to a volume basis by use of Equation 1.1 are often close to values of S_r obtained by extrapolation, but sometimes the values are significantly different. Approximations of laboratory field capacity also may be obtained by allowing samples to reach equilibrium at an arbitrarily high suction, for example, in a centrifuge or in a capillary pressure cell as described in Section 2.5.

2.3.5 Measurement of capillary pressure as a function of saturation - Capillary pressure as a function of S is probably the most important functional relationship in respect to the mechanics of heterogeneous fluids in porous media. Apparently, it was the first functional relationship relating to mixed fluids in porous media to be measured.

Five methods of determining $p_c(S)$ are known, but there are innumerable variations of each. the methods are:

(1) Long Column - A long column of porous medium is allowed to reach equilibrium (with a source of wetting fluid at its base) in the Earth's gravitational field. Saturation is

determined on samples taken at particular elevations after the column reaches equilibrium. The value of p_c at corresponding elevations is determined from Equation 2.5.

According to Collins (1961) this was the first method employed. It is, of course, applicable only for unconsolidated materials which can be packed into a long column. It cannot be used for either undisturbed rocks or soils, and it is useful for determining $p_c(S)$ at relatively small values of p_c only. Furthermore, the time for equilibrium may be very long and uncertain.

(2) Centrifuge - A short column of initially saturated porous material (at most, a few centimeters in length) is placed with its long axis horizontal in a centrifuge as shown in Figure 2-16. The centrifuge is run at some fixed angular velocity until the wetting fluid contained in the sample has reached equilibrium with the centrifugal force imposed by the rotation. A centrifugal force of $g \times 10^3$ is often used.

In most cases, a porous plug is placed at the outer end of the sample during rotation, so that when the rotation is stopped and the sample quickly removed, it will be at some arbitrarily high value of p_c. This procedure is usually used only to obtain an approximation of S_r called the moisture equivalent. An early use of the centrifuge for this purpose was reported by Briggs and McLane (1907).

Another version is to use the centrifuge without the porous plug at the outer boundary of the sample, so that a range of p_c is obtained, from 0 at the outer end to a large value at the inner boundary of the sample. Measurement of S at points along the axis of the sample are obtained during rotation using gamma radiation attenuation. An analysis of the forces acting on the fluids and the distribution of fluids along the axis of the sample during rotation are presented in Section 2.5.

A centrifuge has been used mostly for obtaining $p_c(S)$ on a drainage cycle. This is because it is difficult to connect the sample to a source of wetting fluid during rotation. A method of overcoming this difficulty and obtaining $p_c(S)$ data on an imbibition cycle has been reported by Szabo (1974).

(3) Vapor pressure - A sample of porous material with a known wetting phase content is allowed to equilibrate with the atmosphere inside a closed container. It is limited to cases

for which the non-wetting phase is gas, usually water-air systems. The value of p_c at equilibrium is correlated with the vapor pressure of the wetting phase at equilibrium. Versions of the procedure are available for both the wetting and drying cycles.

The method is useful primarily for studying p(S) beyond the range in which p_c has the usual significance. This is because vapor pressure is very insensitive to pressure changes over the range in which p_c can be related to pore dimensions by Equation 2.12. It becomes more sensitive to changes in liquid saturation when most of the liquid is under the influence of strong adsorptive forces. In this case, the value of p determined is an apparent pressure as explained in Section 2.1.3.

(4) Pressure cell - A sample of porous medium is placed in contact with another fully saturated porous medium having an entry pressure such that it will not desaturate at any p_c imposed during the experiment. An illustration of one such device is shown in Figure 2-15. The porous medium under the sample is called a capillary barrier or a semi-permeable barrier in petroleum literature. The term "semi-permeable" used in this context implies that during the experiment, it will permit the passage of the wetting phase but not the non-wetting phase because of its large entry pressure.

In the soils literature, a capillary barrier is usually identified by a name alluding to the material from which it is made, e.g., a pressure plate (porous ceramic or fritted glass), or pressure membrane (a thin sheet of porous plastic), etc. In any case, the material must have a p_c large enough that it will not permit the breakthrough of the non-wetting phase over the range of p_c needed to desaturate the sample to the degree desired.

The sample is placed on one side of the barrier (usually in a confined chamber surrounded by the non-wetting phase), and the wetting phase is on the opposite side of the container in such a way that its pressure can be controlled. The p_c at the surface of the barrier in contact with the sample is adjusted either by adjusting the wetting phase pressure in the barrier or by adjusting the non-wetting phase pressure in the sample chamber. When the sample has reached an equilibrium saturation at the imposed p_c, its saturation is determined by weighing the sample or by measuring the amount of wetting fluid that was discharged or imbibed during the increment of p_c imposed.

The first use of such a device, apparently, was reported by Willard Gardner et al. (1922). They called their device a "capillary potentiometer" and used it to obtain an empirical relationship for $p_c(S)$ for a soil on the Utah Agricultural Experiment Station at Logan, Utah. The first plots of $p_c(S)$ analogous to those shown in Figure 2-4 were presented by L.A. Richards (1928). In that paper, Richards described the use of a similar device (now called a tensiometer) for measuring the suction of a soil in the field. He also described plans to determine the conductivity of partially saturated soils using an analogous device, which he eventually succeeded in doing as explained in Chapter III. A more detailed discussion of the use of tensiometers for measuring the "capillary tension" in soil water was presented by Richards and Gardner in 1936.

Since the early work of Gardner and Richards, the pressure cell has become the most common device for determining $p_c(S)$ by petroleum as well as soil scientists. Innumerable variations of the technique have been used. One variation designed by A. T. Corey was used by White et al. (1970). With this variation, the sample is not enclosed and evaporation is allowed to proceed continuously. The liquid on the outflow end of the capillary barrier is connected to a capillary tube that contains a liquid-air interface. The tube is connected to a vacuum controller to permit reducing the wetting phase pressure in increments. After each reduction, the wetting liquid moves outward in the tube until evaporation from the sample causes the liquid-air interface to retreat. At the instant the interface begins to retreat, the sample is weighed to determine its saturation. The sample is then returned to the cell, the wetting fluid pressure is reduced by another increment and the process is repeated. The advantage of the latter technique is that a balance of pressure is achieved much more quickly when evaporation is permitted, especially at lower values of saturation.

(5) Brooks method - A method has been devised by R. H. Brooks (1980) which has great utility in respect to obtaining $p_c(S)$ on an imbibition cycle, which is difficult to accomplish by any other method. With the Brooks system, a carefully metered quantity of wetting fluid is added to a porous sample. After a short time (necessary for the wetting-fluid pressure to stabilize) the fluid pressure is recorded using a capillary barrier in contact with the soil and connected to a null pressure transducer. The saturation is determined from the known quantity of water absorbed. The advantage of this procedure is that the sample is not required to imbibe fluid from a low pressure source through a capillary barrier. The problem of maintaining excellent contact with the barrier is removed and

equilibrium is established in a small fraction of the time required with a pressure cell.

2.3.6 Empirical representations of $P_c(S)$ - Brooks and Corey (1966) plotted $\ell n\ S_e$ as a function of $\ell n\ p_c$. They found that if they omitted data for $p_c < p_e$, the data plotted on a straight line as shown in Figure 2-6, provided that a suitable value of S_r is used to compute S_e from measured values of S. In fact, Brooks and Corey determined S_r as that value of S which provided the most nearly linear curve. It was found that the shape of the curve is very sensitive to the value of S_r selected. An extrapolation of the linear curve provides a sensitive evaluation of p_d.

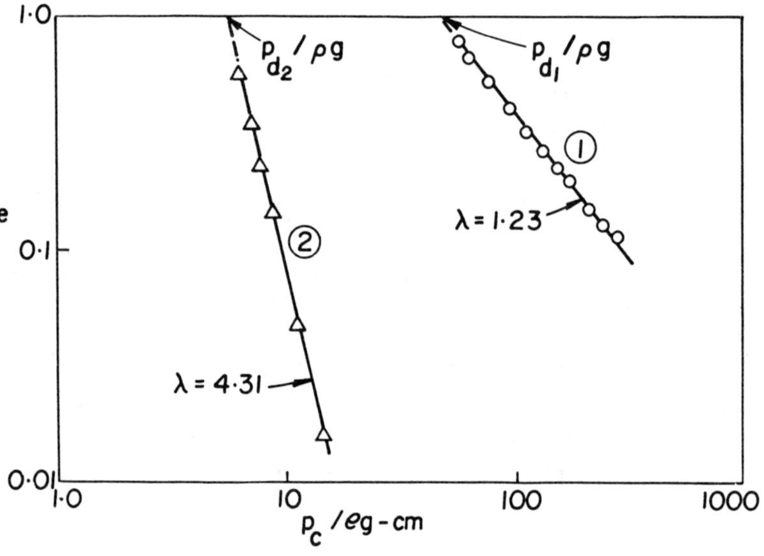

Figure 2-6. Effective saturation as a function of capillary pressure.

As a result of their observation that data for all relatively homogeneous and isotropic samples provided curves as shown in Figure 2-6, Brooks and Corey suggested the empirical relationship

$$S_e \cong \left(\frac{p_d}{p_c}\right)^{\lambda} \qquad 2.9$$

for $p_c > p_d$ and $S > S_r$. Originally Equation 2.9 was suggested for use only on a drainage cycle, but Su and Brooks (1975) have applied an analogous relationship (Equation 2.11) for the imbibition cycle as well. In the latter case, a different interpretation is given to the parameter p_d.

Brooks and Corey found that for typical porous media, λ is about 2. Soils with well-developed structure have values of λ less than 2 and sands normally have values of λ greater than 2, sometimes as large as 5 or more.

Equation 2.9 does not represent the measured data for values of $p_c < p_d$. Other empirical expresions have been suggested for this purpose. One such expression, suggested by Laliberte (1969), is

$$S_e = 0.5(1-\text{erf}\xi) \qquad S_e \leqq 0.5$$

$$S_e = 0.5(1+\text{erf}\xi) \qquad S_e \geqq 0.5 \quad . \tag{2.10}$$

In Equation 2.10, ξ is a function of p_c describing the pore-size distribution.

Another such expression, suggested by Su and Brooks (1975), is given by

$$p_c = p_e\left[\frac{S-S_r}{a}\right]^{-m}\left[\frac{1-S}{b}\right]^{bm/a} \tag{2.11}$$

in which a, b and m are constants. The constant m is roughly equivalent to $1/\lambda$ in the Brooks-Corey relationship. Su and Brooks have described the physical interpretation of the constants a and b in their paper. Equation 2.11, like Equation 2.10, is designed to fit data for all values of p_c.

White et al. (1970) have also presented a semi-analytical relationship for the entire curve of $p_c(S)$, but in their paper, they pointed out that the values of $p_c < p_e$ are a function of the external sample geometry and not the pore geometry. According to their theory, the desaturation that occurs at $p_c < p_e$ proceeds only from the boundaries of samples, and consequently is a function of the ratio of external boundary area to the volume of the sample.

Occasionally, samples are found with a $p_c(S)$ function unlike that predicted by any of the empirical models. These are often found to be obviously non-homogeneous materials, for example, layered rocks. Curves of $\ell n\ S_e$ as a function of $\ell n\ p_c$ often consist of two or more straight-line portions offset by an increment of $\ell n\ p_c$ and connected by an abrupt transition.

A similar result is produced by samples which consist of a sieve fraction of soil aggregates or a sieve fraction of crushed (but not pulverized) sandstone rock. In this case, the two-staged curve (sometimes called bimodal) undoubtedly results from a discontinuity in pore size between the primary and secondary pore space. However, this type of bimodal behavior has not been observed with undisturbed soil samples.

2.4 PORE-SIZE DISTRIBUTION

From a force balance similar to that used in Section 1.6.1 to obtain Equation 1.5, it is possible to obtain $\bar{p}_c$ for an interface across an irregular shaped space such as may exist in soil pores. In general, the line of contact of an interface with the internal solid surfaces of porous media is not circular. However, because of the tendency of interfaces to reach a minimum area consistent with the saturation, the line of contact tends to lie nearly in a plane.

In the following analysis, sections of interfaces are considered which are concave toward the non-wetting fluid and which possess orthogonal radii of curvature of the same sign, as well as having lines of contact that lie more or less in a plane. Adsorbed films and pendular rings are excluded. A sketch of a portion of pore space (across which several such sections might be positioned) is illustrated in Figure 2-7.

A balance of forces across a particular section indicates that

$$\bar{p}_c A \cong \sigma \cos \alpha \ wp$$

where wp designates the wetted perimeter, that is, the length of the line of contact. The area A represents the area (of the plane passing through wp) which is enclosed by wp. The force balance presumedly would be exact if the wp under consideration actually lies exactly in one plane, if the contact angle α is constant over wp and if σ is constant over the section.

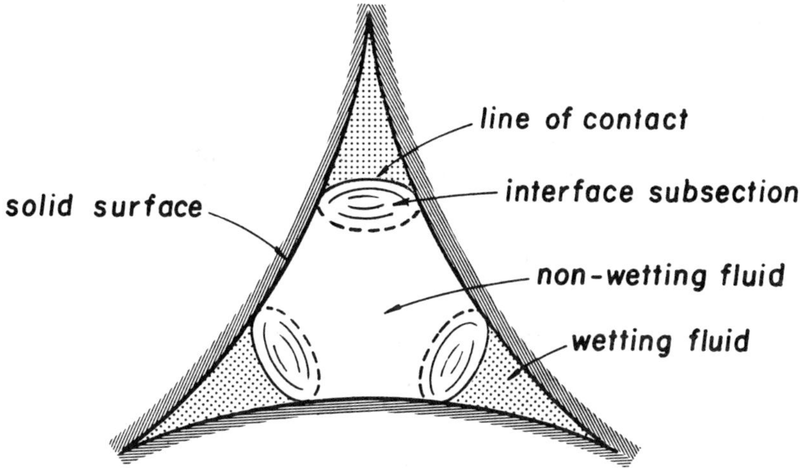

Figure 2-7. Cross section of pore space with several interface sections.

Rearranging gives

$$\frac{A}{wp} \cong \frac{\sigma \cos \alpha}{\bar{p}_c} \quad . \tag{2.12}$$

The quantity A/wp has the dimension of length and can be used to characterize the "size" of the section of pore space across which interfaces are positioned. Note, that this concept is consistent with the definition of pore size given in Section 1.4.3. It is also consistent with the concept of hydraulic radius. However, it should be observed that the geometry of a particular pore region might not be characterized by the value of p_c in surrounding regions if the non-wetting phase is blocked from entering the region by smaller pore sizes at all points on the boundary of the region.

The following analysis assumes that practically all portions of the pore space with pore sizes larger than that given by Equation 2.12 have access to the non-wetting phase. This may appear to be a severe restriction, but evidence presented by White et al. (1972), and Corey and Brooks (1975) imply that the condition is satisfied for practical purposes at saturations smaller than that corresponding to the inflection point occurring at the capillary pressure p_e. However, pore sizes represented by $p_c < p_e$ must be characterized by extrapolation as is explained later.

A volume element of the porous medium is considered which is large enough to be representative of a particular sample but is small enough that p_c can be represented by a single value when the fluids are in a static condition on the drainage cycle. This implies that the elevation difference within the element is negligible. Within such an element, there usually are many sections of the interface. The quantity A/wp should be the same for each, regardless of the orientation of A. Furthermore, if the medium is isotropic, the orientations of A for each section should be random, that is, there should be no preferred orientation.

If p_c is increased within the reference element, the value of A/wp will decrease. In the process, a portion of the pore space will lose its wetting phase, that is, S will decrease. The increment of S is a measure of the fraction of pore space characterized by the corresponding increment of A/wp. The ratio of change in S to change in A/wp is dependent upon the frequency of pore sizes which have values of A/wp within the increment of p_c considered.

Equation 2.12 indicates that if $\sigma \cos \alpha$ is essentially constant for a particular fluid system, A/wp should be proportional to $1/p_c$. The quantity $1/p_c$, therefore, should be a measure of the largest pore "size" (in that part of the pore space containing the wetting phase) at a particular value of p_c. Similarly, a plot of S as a function of p_d/p_c should provide an indication of the distribution of "sizes" characterizing the pore space. Figure 2-8 presents contrasting examples of this kind of plot.

In Figure 2-8, S_e rather than S has been plotted as a function of p_d/p_c. By normalizing the ordinate and abscissa in this way, data for any medium can be plotted on the same graph without changing the scale. Furthermore, for media with similar pore-size distributions, the data should plot on the same curve. In any case, the end points of all curves should be the same, that is, 0 and 1.0. This is because both S_r and p_d are chosen by an extrapolation process which forces the curves to behave in this way.

It should be observed that p_d is somewhat smaller than p_e. However, actual data for $p_c < p_e$ are disregarded in the extrapolation process. The rationale for this is that for $p_c < p_e$, all regions of the pore space may not have access to the non-wetting phase, so that the geometry of such regions is

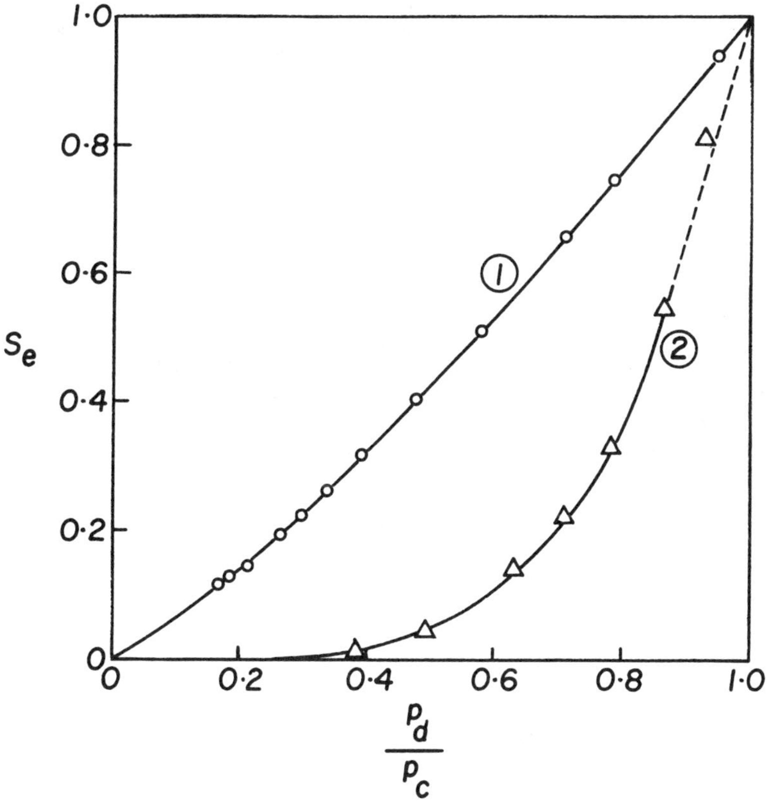

Figure 2-8. Effective saturation as a function of p_d/p_c.

not characterized by $p_c(S)$ data for $p_c < p_e$. Pore sizes larger than that corresponding to p_e are assumed to be determined by the extrapolation process, the largest being characterized by p_d.

Data for high values of p_c, where S approaches S_r, are also disregarded. The reason for this is that a significant portion of the water which is removed at such saturations may be associated with surface films and pendular rings. There is no reason to suppose that Equation 2.12 would be valid in this situation, or that the pore geometry would be described by $p_c(S)$ where S approaches S_r.

Curves of $p_c(S)$ have been used by many investigators to characterize pore-size distribution in porous media. Perhaps the first to suggest this were Childs and Collis-George (1948) and Purcell (1949). However, none of the investigators employing this principle have pointed out the apparent fallacy of including data for $p_c < p_e$ in the characterization of pore geometry.

Many investigators, Purcell being among the first, have used data for $p_c(S)$ obtained by a mercury injection procedure. In the latter case, mercury is injected as a non-wetting phase into a porous sample which has been first evacuated, the wetting-phase space being empty. The value of p_c is taken as the pressure required to inject a given increment of mercury into the pore space and S is calculated from the volume of fluid injected. The method of mercury injection has the advantage of speed, but it does not evaluate any changes in pore size that might result from the reaction of porous media to fluids of the prototype system.

The first step in plotting curves of the type shown in Figure 2-8, after the data are measured, is to determine values of S_r and p_d. One way of accomplishing this is as follows:

(1) The data for $p_c(S)$ are plotted as shown in Figure 2-4.

(2) Estimations of p_e, S_r and p_d are made which are designated as p_e', S_r' and p_d'.

(3) Data representing $p_c < p_e'$ and $S \leqq S_r'$ are disregarded.

(4) Using the remainder of the data, a plot of $(p_d'/p_c)^2$ as a function of S is made as shown in Figure 2-9. For typical pore-size distributions, the plot is roughly linear. This is because an average value of λ [see Equation 2.10] is about 2.

(5) The plot is extrapolated to the abscissa to obtain S_r.

(6) The plot is extrapolated to the ordinate to obtain the intercept I.

(7) p_d is calculated from

$$p_d = p_d'/\sqrt{I} \quad .$$

(8) The improved values of S_r and p_d are used to calculate S_e and corresponding values of (p_d/d_c) so that curves of the type shown in Figure 2-8 can be plotted.

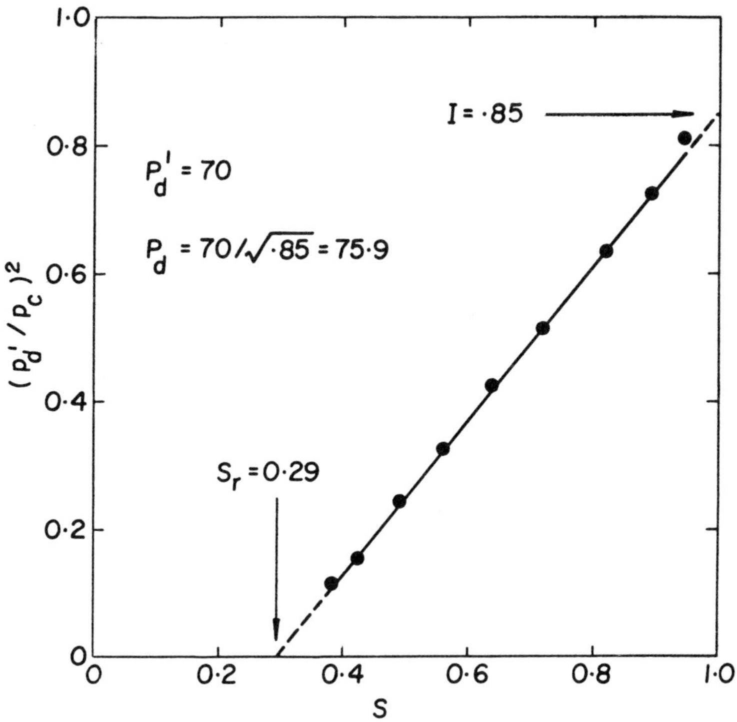

Figure 2-9. Graphical procedure for estimating S_r and p_d.

More sensitive extrapolation procedures for obtaining S_r and p_d can be devised. One such method is presented in Section 2.3.6.

Another method of illustrating the "size" distribution of pore space is by plotting $dS/d(\frac{p_d}{p_c})$ as a function of p_d/p_c. Examples of such plots are shown in Figure 2-10, Curves 1 and 2 representing the same measured data as Curves 1 and 2 in Figure 2-8.

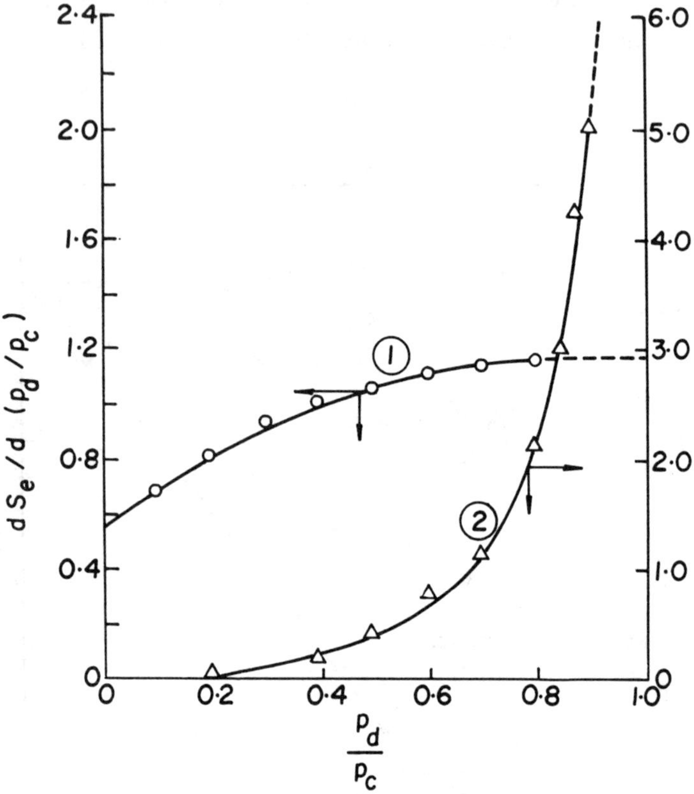

Figure 2-10. Frequency distribution of pore-size as a function of p_d/p_c.

The plots shown in Figure 2-10 can be regarded as frequency distribution curves because they indicate the rate of change of S with respect to a function proportional to pore size. Note that the area under such curves is 1.0.

A medium having a wider range of pore size, e.g., Sample 1, has a curve which is less sharply peaked and which has its area spread over a wider range of values of p_d/p_c than is the case for Sample 2. According to White et al. (1972), the highest value of the ordinate actually measured occurs at the entry pressure p_e. Undoubtedly, this is because the non-wetting phase cannot reach all regions of a medium until p_e

is reached. Consequently, that part of the curves representing $p_c < p_e$ is shown with dotted lines.

2.4.1 Pore-size distribution index - Brooks and Corey used the parameter λ as shown in Figure 2-6 and Equation 2.9 as an index of pore-size distribution. They reasoned that for media having a uniform pore-size, the index would be a large number which theoretically could approach infinity. On the other hand, media with a very wide range of pore sizes should have a small value of λ which theoretically could approach zero. They found that for typical porous media, the usual value of λ is about 2.

In the case of naturally occurring sand deposits, λ is often about 5 or 6, especially if the material is thoroughly mixed and densely packed. Well-aggregated soils in an undisturbed state sometimes have $\lambda < 1$.

2.4.2 Factors affecting pore-size distributions - The effect of grain-size distribution, structure and mixing on pore-size distribution is mentioned in Section 1.4.4. Of these, the grain-size distribution is probably the least important. There is a tendency, however, for finer materials to have smaller values of λ.

All sands, regardless of grain-size distribution can be made to have a very uniform pore-size (high λ) by thorough mixing and dense packing. The mixing and packing can be accomplished by dropping the sand through at least two screens a few centimeters apart as suggested by Wygal (1963). Even undisturbed sands found in natural deposits usually have relatively high values of λ, say 4 or 5.

Finer materials usually have somewhat smaller values of λ, but in this case also, the value of λ can be increased by thorough mixing and dense packing. Laliberte and Brooks (1967) experimentally determined the relationship between ϕ and λ for several soils and sands and found that λ increases as ϕ decreases. Evidently, the decrease in ϕ reduces the range of pore sizes as might be expected.

Clearly, structure increases the range of pore sizes and decreases λ, but the effect of shape of sand grains is not known. It is not known, for example, whether a freshly crushed deposit of rock fragments would have a significantly smaller value of λ than sand which is taken from a stream bed.

It would be expected that porous rocks would have a characteristic range of λ values depending upon their geological classification. The degree and type of cementation might also have an effect. Apparently such relationships have not been studied systematically, however.

2.5 DISTRIBUTION OF FLUIDS IN STATIC SYSTEMS

If the pressure history of a static system is known and the relationship $p_c(S)$ can be measured, for each variety of medium contained in the system, it is possible to deduce the distribution of fluids. This is accomplished by an application of Equation 2.5. Examples of static fluid distributions are described, some relating to field systems and some relating to laboratory devices, in the following sections.

2.5.1 A soil-water system - As a first example of the application of Equation 2.5, a soil profile is considered which has been thoroughly soaked following a rain or irrigation. If such a soil is protected from evaporation and from large temperature fluctuations, the soil-water system will eventually approach a state of equilibrium. If there is a water table at some depth below the surface, the fluid system will be in equilibrium with the water table and the water table can be used as a datum from which to measure the elevation and pressure.

In the case of a soil in the field, a pseudo static situation can be expected only in a fallow soil, where plants are not present to remove the water continuously and where enough surface soil has dried to protect the remainder from significant evaporation. In such cases, the water some distance below the dry layer may approach a static distribution.

Field soils never reach a condition of complete saturation, however, either above or immediately below the water table. This is because some entrapped gases are always present. Usually about 8-20 percent of the pore space is occupied by gases even where the water saturation is a maximum. Under such circumstances a saturation distribution similar to that shown in Figure 2-11 might be expected if the soil profile is homogeneous.

In Figure 2-11, h represents an elevation measured above the water table, that is, the locus of points at which the water is at atmospheric pressure. Note that the water content both immediately above as well as below the water table is at some maximum value S_m which is determined by the saturation at which gas ceases to be interconnected on the wetting cycle following rains or irrigations. It is not known at this time to what depths an entrapped gas might extend, but it does extend at least a meter below a water table.

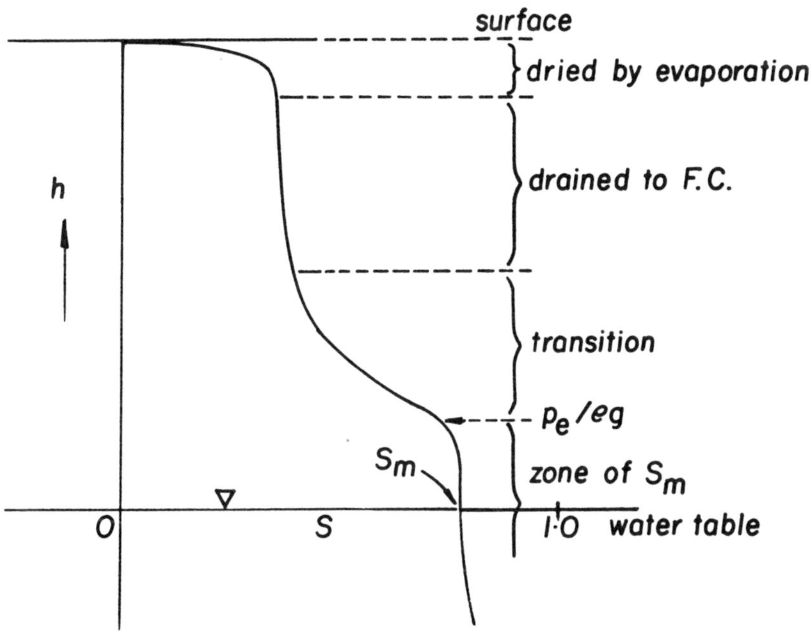

Figure 2-11. Distribution of pseudo static water in a homogeneous soil profile with water table.

The relationship between Figures 2-11 and 2-4 also should be observed. Note that part of the water distribution curve for the profile (below the zone of evaporation and above the water table) is similar to the relationship shown in Figure 2-4 except that the ordinate p_c has been replaced by h in conformance with Equation 2.6; that is, h is $p_c/\rho_w g$.

A second example of a "pseudo" static water distribution is shown in Figure 2-12. In this case there is no water table, a layer of dry soil being beneath as well as above the moist soil.

The abbreviation PWP in Figure 2-12 refers to the <u>permanent wilting point</u>, a term used by soil scientists to designate that water content which remains after plants growing in the soil have removed as much water as is possible. PWP usually is about half of FC but greater than the water content of air-dried soil. The distributions illustrated in Figures 2-11 and 2-12 do not represent conditions of true static equilibrium, of course, because the presence of sharp transitions from wet to dry soil imply a pressure gradient that is certain to cause some flow. The enormous resistance in regions of dry soil, however, makes possible a pseudo static condition in the wet soil. This situation is more easily explained by theory discussed in Chapter IV.

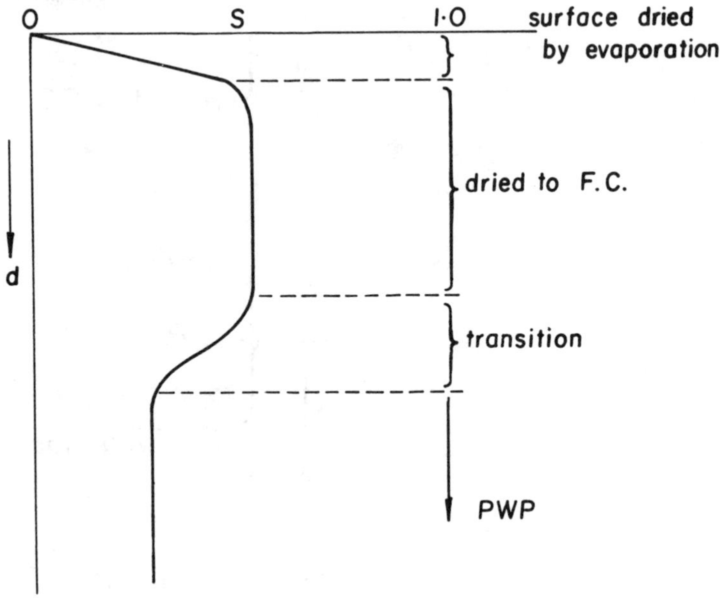

Figure 2-12. Distribution of pseudo static water in a homogeneous soil without a water table.

Actual soil profiles often consist of layers having contrasting properties in respect to the $p_c(S)$ relationship. To determine a water distribution curve for such a case, it is necessary to obtain a $p_c(S)$ relationship on the appropriate cycle for each layer in the sequence. The distribution of p_c with elevation when a water table exists is not affected by the presence of contrasting strata, provided the system is truly static. However, in this case, even an infinitesimal flow may change the distribution of p_c drastically. Consequently, the distribution of fluids in non-homogeneous media is better discussed in the context of theory presented in Chapter IV.

A question that frequently needs to be answered, relative to a soil-water system as illustrated in Figure 2-11, is how much water will drain from the soil if the water table is lowered to another static position. In some literature, the volume of water per unit surface area that will drain if the water table is lowered by a unit increment is called specific yield. In this text specific yield S_y is defined by

$$S_y \equiv \frac{dV_d}{dD} \qquad 2.13$$

in which V_d is the volume drained per unit surface area and D is the depth to the water table from the dry surface layer.

Equation 2.9 provides a convenient tool for evaluating this derivative. In laboratory samples, Equation 2.9 is not completely valid for $p_c < p_e$, but the boundary effect that causes the initial desaturation is not expected to have a significant effect in the massive soil volume. This is because air can replace water only by moving downward from the desaturated zone and not from lateral boundaries as would be the case with small samples.

For the field soil, however, the value of S is not 1.0 at any point, even below the water table where air is trapped. Therefore, S_e is redefined for this case as

$$S_e \equiv \frac{S - S_r}{S_m - S_r}$$

in which S_m is the maximum field saturation. Also, Equation 2.9 is modified in terms of Equation 2.5 as

$$S_e \equiv \left(\frac{h_d}{h}\right)^{\lambda}$$

for $h > h_d$, in which h_d corresponds to an elevation at which p_c is p_d. The meaning of p_d may be modified slightly because of the fact that the maximum water saturation is S_m rather than 1.0.

The volume drained below the dry layer is given by

$$V_d = \phi_e \int_0^C (1 - S_e)dh \quad ,$$

in which ϕ_e is here defined as

$$\phi_e \equiv \phi(S_m - S_r) \quad .$$

Expressing S_e in terms of h gives

$$V_d = \phi_e \int_{h_d}^{D} [1 - (\frac{h_d}{h})^\lambda] \, dh \quad .$$

Since the derivative of an integral with respect to a variable upper limit (the lower limit being constant) is the integrand evaluated at the upper limit,

$$S_y = \phi_e \, [1 - (\frac{h_d}{D})^\lambda] \qquad 2.14$$

for $D \geqq h_d$, and for $D < h_d$,

$$S_y = 0 \quad .$$

Note that for $D >> h_d$,

$$S_y \cong \phi_e \quad .$$

It is often assumed by groundwater and drainge engineers that S_y is ϕ_e, which they call the drainable porosity. However, the value of D at which $(h_d/D)^\lambda$ becomes negligible compared to 1 depends upon the value of λ. For sandy soils which have values of λ of about 4 or 5, ϕ_e is a good approximation of S_y when D is only about $2h_d$.

On the other hand, soils with considerable structure often have values of λ less than 1.0, in which case, ϕ_e is not a good approximation of S_y unless the water table is much deeper than $2h_d$. It often happens, especially in respect to drainage design, that engineers make a larger error than is realized by regarding S_y as a constant. This fact has been pointed out by Duke (1972) who is responsible for the derivation of Equation 2.14. An excellent discussion of the concept of specific yield has been presented also by Bear (1972).

2.5.2 Petroleum reservoir - An oil-bearing rock formation is another example of a system which sometimes may contain practically static fluids. In such cases, the system has fluid distributions somewhat analogous to those of a water-air system in a soil.

A typical petroleum reservoir is a stratum of porous rock confined by relatively impermeable rock. Usually the petroleum fluids (oil, or oil and gas) occupy only a small fraction of the porous stratum, and even there, they are mixed with the brine that inevitably saturates the remaining and larger portion of the stratum. If the fluids are static, the oil and gas are in the higher part of the stratum, with gas at the apex.

Fluid carbon compounds, which make up the oil and gas, are thought to have been formed in mud buried in saline marine deposits under reduced conditions. Consequently, there was insufficient oxygen to oxidize the carbon compounds. Later the mud underwent metamorphosis and became shale. In this process, the heat and pressure forced much of the oil and gas out into adjacent strata, in some cases aquifers, where it could accumulate and become mobile.

The petroleum fluids were able to accumulate, however, only in formations that were sealed. Such formations were necessarily confined between adjacent impermeable strata, and capped at the highest point. Owing to their buoyancy, the density of most petroleum fluids being less than that of brine which originally fully occupied the sandstone, the petroleum fluids tended to migrate upward until confined by the cap rock.

Often a petroleum body is found in a formation at the apex of an anticline, as illustrated in Figure 2-13, or where the oil-bearing aquifer slopes upward to a point where it is blocked by a fault. In some cases, discussed in Chapter IV, the cap rock is also porous but with an entry pressure (oil into brine) sufficient to prevent further upward migration. Usually, the latter type of cap rock is sufficient to provide a significant accumulation only under dynamic conditions in which brine is flowing down slope.

Fluids in petroleum reservoirs are usually at elevated temperatures and at pressures much greater than atmospheric. Conequently, there is nothing completely analogous to a water table in a petroleum reservoir. Also, the carbon compounds often exist in two phases, i.e., liquid and gas. The gas occupies the apex of the petroleum body as shown in Figure 2-13.

Among the first questions to be considered is how much oil is recoverable and where is the best place to drill wells to accomplish the most efficient recovery. As the oil is removed the brine will migrate upward and the gas downward until eventually oil production becomes unprofitable either because too much brine is produced along with too little oil or the gas breaks through to the wells. After gas breakthrough, the reservoir pressure may fall precipitously and eliminate the energy available for recovering the oil.

To answer all of the questions posed obviously requires consideration of the dynamics of the system. In this section, only the static distribution of fluids in the virgin reservoir

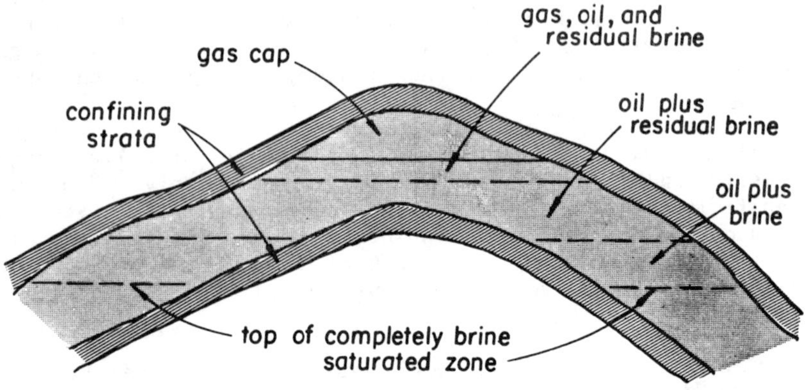

Figure 2-13. Schematic of petroleum reservoir with static fluids.

is considered. To determine this, the distribution of p_c must first be computed. The lower edge of the hydrocarbon region may be used as a starting point for the calculation of p_c. The data needed are:

(1) the densities of each of the fluid phases, including brine,
(2) p_e for oil into brine,
(3) p_e for gas into oil,
(4) The elevations of the lower edges of the oil and gas bodies, or,
(5) The absolute pressures of the oil, gas and brine at some point or points where these three phases exist,
(6) S_w, S_o and S_g as functions of p_e, two p_c functions being involved, that is, oil into brine and gas into oil.

The value of p_c at the brine-oil interfaces is given by Equation 2.5 where h is the elevation above the bottom of the oil body and p_{c_o} is p_e (oil into brine) for the rock at that position.

Similarly, p_c (gas into oil) in the upper portion of the stratum is given by Equation 2.5 where h is the elevation above the bottom of the gas body and p_{c_o} is p_e (gas into oil) for the rock at that position. The brine, oil and gas distributions can be estimated from $p_c(S)$ curves although a complication exists within that portion of the gas cap where three fluid phases may exist. This is usually a relatively narrow zone, however.

A possible distribution of fluids existing in such a reservoir is shown in Figure 2-14, in which the distribution is with respect to elevation above the bottom of the oil body.

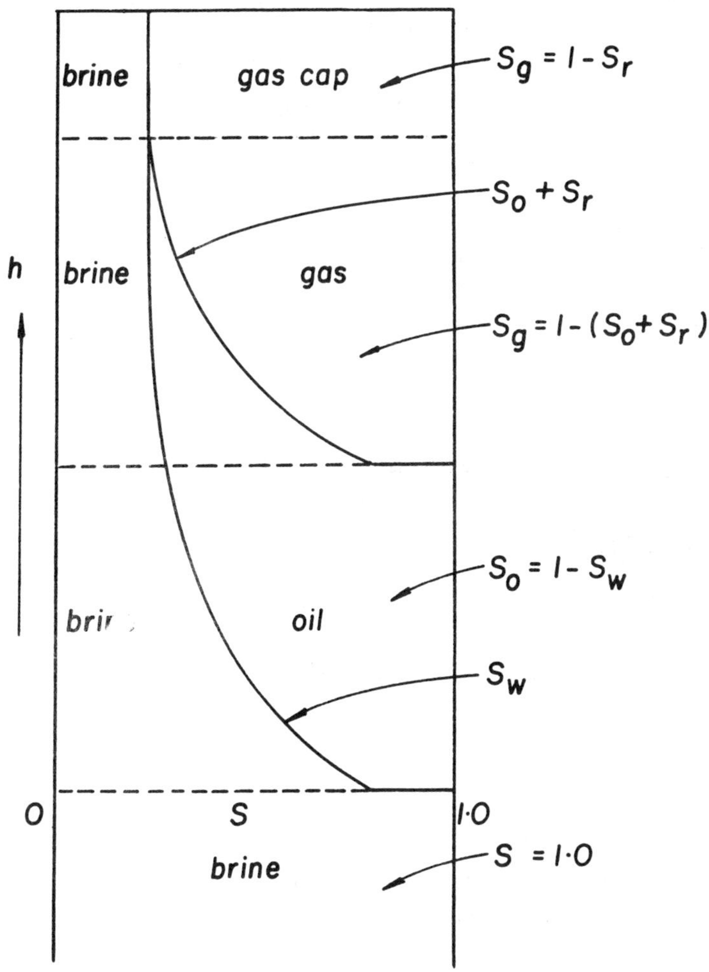

Figure 2-14. Distribution of fluids in a petroleum reservoir.

Complications are caused also by stratification as in the case of soils. Furthermore, a stratum of oil in a state of true equilibrium is rarely found even for virgin reservoirs. Often the brine is found to be flowing slowly in one direction or another, and this has a large effect on the distribution of fluids. In the static case, however, S_w decreases and S_o increases with elevation until S_w becomes S_r. At some elevation, a gas body begins, at which level p_c (gas into oil) is p_e for gas into oil. Above this level, S_o decreases and brine remains at the saturation S_r. At a higher level, the oil body may disappear completely so that only gas and residual brine remain. That part of the reservoir which contains only gas and residual brine is called a gas cap.

2.5.3 Fluids in a porous solid in equilibrium with a capillary barrier

- A sketch of a typical capillary-pressure cell is shown in Figure 2-15. As explained in Section 2.3.5, a pressure cell is used to control p_c at one boundary of a saturated porous medium, that is, at the boundary of a capillary barrier. When a porous sample is placed in contact with the barrier, its saturation will adjust until the p_c of the sample at the plane of contact also reaches the controlled value.

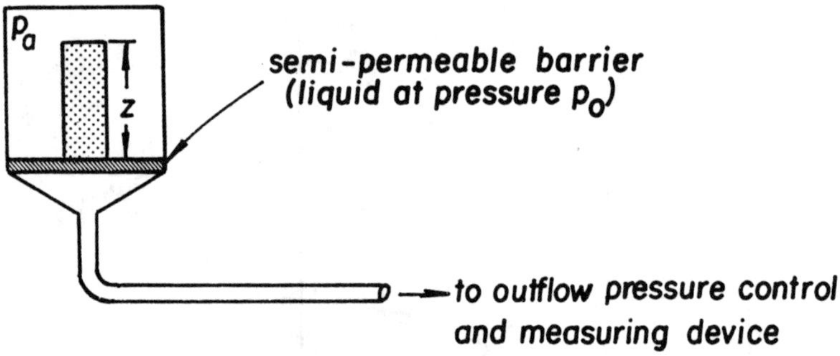

Figure 2-15. Capillary-pressure cell.

In reference to Figure 2-15, if p_o is maintained in the liquid at the top of the barrier and a pressure p_a in the gas within the chamber, the following relationships will hold at equilibrium:

At $z = 0$

$$p_c = p_a - p_o.$$

At any z,

$$p_c = p_a - p_o + \rho_w g z.$$

At $p_o = 0$ and $z = 0$,

$$p_c = p_a .$$

It is necessary, of course, that the temperature and vapor pressure inside the sealed chamber also be allowed to reach equilibrium. For example, if the chamber is connected to an air pressure regulator, provision should be made to minimize diffusion of the wetting-fluid vapor out through the air line to the regulator.

It should also be noted that p_c of the sample at equilibrium is not a constant but varies with elevation according to the term $\rho_w gz$. If the sample height is small, however, this may be of little significance.

2.5.4 Fluids at equilibrium in a porous solid in a centrifuge

- A small sample of porous material (containing a wetting liquid) being rotated about an axis in a centrifuge is sketched in Figure 2-16.

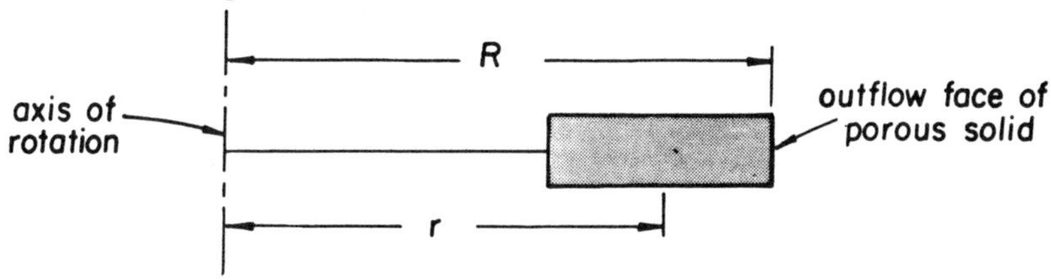

Figure 2-16. Schematic of porous sample in centrifuge.

The condition of the liquid in the sample (after the centrifuge has been rotated for a sufficient time at a constant angular velocity ω) can be analyzed as a problem of statics. The fluid elements no longer move with respect to the solid matrix. At this time, the pressure gradient in the liquid is such as to balance the centrifugal force and gravity. In this case gravity can be ignored because centrifugal force is imposed which has a magnitude of about $g \times 10^3$.

Since the centrifugal force is in the outward direction on fluid elements, the negative pressure gradient must act inward. The balance of forces is given by

$$-\frac{dp_w}{dr} + \rho\omega^2 r = 0 \quad ,$$

where r is a distance outward from the center of rotation. If the density and pressure of air are taken to be zero,

$$dp_c = -\omega^2 \rho r dr \quad .$$

Integrating gives

$$p_c = -\frac{\omega^2}{2}\rho r^2 + c \quad .$$

A consideration of the physics of this system indicates that on the outflow face of the sample, at $r = R$, the liquid pressure is only negligibly greater than the pressure of surrounding air. Thus

$$c \cong \frac{\omega^2}{2}\rho R^2$$

and

$$p_c \cong \frac{\omega^2}{2}\rho(R^2 - r^2) \quad , \qquad 2.15$$

where R is the radius at the outflow end of the sample.

Capillary pressure varies from zero where the radial distance is R to a higher p_c towards the center of rotation, depending upon ω. In practice p_c can be made very large at a short distance from the outflow face. If it is desired to determine S_r for a sample, a porous plug is placed in contact with the outflow face of the sample. This, in effect, moves the outflow face of the system outward from the sample and places the entire sample in the high p_c region. When the rotation of the centrifuge is stopped, negligible redistribution of water in the sample will take place, because the entire sample will be at S_r and will have a negligible conductivity of water.

REFERENCES

Bear, Jacob (1972), Dynamics of Fluids in Porous Media. American Elsevier Publishing Co., Inc. New York. p. 483.

Briggs, L. J., and McLane, J. W. (1907), The moisture equivalent of soils. U.S. Dept. Agr. Bur. Soils, Bulletin 45, 23 p.

Brooks, R. H. (1980), Water retention measurement for soils. J. Irrig. and Drainage Div., ASCE, Vol. 106, No. IR2, pp. 105-112.

Brooks, R. H. and Corey, A. T. (1966), Properties of porous media affecting fluid flow. J. Irrig. and Drainage Div., Proc. ASCE, IR 2, pp. 61-88.

Childs, E. C. and Collis-George, N. (1948), Soil geometry and soil-water equilibria. Disc., Faraday Society, No. 3, pp. 78-85.

Collins, R. E. (1961), Flow of Fluids Through Porous Materials. Reinhold Publishing Corporation, New York, 270 p.

Corey, A. T. (1954), The interrelation between gas and oil relative permeabilities. Producer's Monthly, Vol. 19, pp. 38-41.

Corey, A. T. and Brooks, R. H. (1975), Drainage characteristics of soils. Soil Science Society of American Proceedings, Vol. 39, No. 2, pp. 251-255.

Duke, H. R. (1972), Capillary properties of soils - influence upon specific yield. Transactions, ASAE, Vol. 15, No. 4, pp. 688-691.

Edlefsen, N. E., and Anderson, A. B. C. (1943), Thermodynamics of soil moisture. Hilgardia, Vol. 15, No. 2, 298 p.

Gardner, W., Israelsen, O. W., Edlefsen, N. E., and Clyde, H. (1922), The capillary potential function and its relation to irrigation practice. Physical Review, second series, July-December, p. 196.

Laliberte, G. E. (1969), A mathematical function for describing capillary pressure-desaturation data. Bulletin of the International Association of Scientific Hydrology, Vol. XIV, 2, pp. 131-149.

Laliberte, G. E. and Brooks, R. H. (1967), Hydraulic properties of disturbed soil materials affected by porosity. Proc. SSSA, Vol. 31, pp. 451-454.

Miller, E. E. and Klute, A. (1967), The dynamics of soil water, Part I - Mechanical forces, In Irrigation of Agricultural Lands, R. M. Hagan, et al. (eds.), Chapter 13, pp. 209-244.

Peters, D. B. (1965), Water availability. Chapter 19, Methods of soil analysis, Agronomy, No. 9, Part 1, ASA, pp. 279-285.

Purcell, W. R. (1949), Capillary pressures--their measurement using mercury and the calculation of permeability therefrom. Journal of Petroleum Technology, Vol. 1, No. 2, pp. 39-46.

Richards, L. A. (1928), The usefulness of capillary potential to soil-moisture and plant investigators. Journal of Agricultural Research, Vol. 37, No. 1, pp. 719-742.

Richards, L. A., and Gardner, W. (1936), Tensiometers for measuring the capillary tension of soil water. Journal of the Amer. Soc. of Agronomy, Vol. 28, pp. 352-358.

Richardson, J. G. (1961), Flow through porous media. Handbook of Fluid Dynamics, Section 16, edited by V. I. Streeter, McGraw-Hill Book Co., Inc., New York, pp. 16-3 through 16-112.

Shaw, C. F. (1927), The normal moisture capacity of soils. Soil Science, Vol. 23, pp. 303-317.

Su, Charles and Brooks, R. H. (1975), Soil hydraulic properties from infiltration tests, Watershed Management Proceedings, Irrig. and Drain. Div., ASCE, Logan, Utah, August 11-13, pp. 516-542.

Topp, G. C. (1969), Soil-water hysteresis measured in a sandy loam and compared with the hysteretic domain model. Proc. SSSA, Vol. 33, No. 5, pp. 645-651.

White, N. F., Duke, H. R., Sunada, D. K., and Corey, A. T. (1970), Physics of desaturation in porous materials. Journal of the Irrig. and Drain. Div., Proc. ASCE, IR 2, pp. 165-191.

White, N. F., Sunada, D. K., Duke, H. R., and Corey, A. T. (1972), Boundary effects in desaturation of porous media. Soil Science, Vol. 113, No. 1, pp. 7-12.

Wygal, R. J. (1963), Construction of models that simulate oil reservoirs. Society of Petroleum Engineers Journal, December, pp. 281-286.

Zemansky, M. W. (1943), Heat and Thermodynamics, Chapter III, Section 3.2, second edition. McGraw-Hill, New York, pp. 21-25.

PROBLEMS AND STUDY QUESTIONS

1. Consider a sealed tube of soil partially saturated with water. A small temperature differential is maintained continuously across the ends of the tube. (a) Will the tube eventually reach a condition of thermal equilibrium? (b) Will it reach a condition of mechanical equilibrium? Give reasons for each answer.

2. Consider the same situation as described in question 1 except that the soil is replaced by a hypothetical medium having pores small enough to prevent bulk flow (too small to hold fluid elements). Answer question (a) and (b) above and give reasons.

3. Explain why pressure p is treated as a scalar quantity, whereas $\underline{\sigma}_c$, the conservative surface stress, is treated as a vector.

4. Explain why (when balancing forces on a reference element of fluid) it is not necessary to consider surface forces produced at solid-fluid boundaries, whereas for macroscopic reference elements, this is necessary.

5. Give a possible reason why soil water at a negative gauge pressure of 5 atmospheres does not cavitate, whereas water in pipes usually cavitates at a negative pressure of 1 atmosphere.

6. What is the meaning, if any, of capillary pressure in a soil at a point a few centimeters above a water table? How could this capillary pressure be measured?

7. Consider two capillary tubes, one having a circular cross section, and the other a square cross section. If both tubes have the same cross-sectional area, which tube is likely to have the higher capillary rise? Explain. Answer the same question in respect to a capillary tube having a rectangular cross section, with one side being twice the dimension of the other.

8. Consider a straight capillary tube of uniform cross-sectional area but with varying cross-sectional shape. Would such a tube be expected to exhibit hysteresis in an analogous way to the tubes illustrated in Figure 2-5? Explain.

9. It is supposed that the small amount of desaturation occurring at $p_c < p_e$ is due to retreating interfaces at the sample boundary. Describe a possible experiment to provide evidence of this premise.

10. Would the correlation between S_r (found by extrapolation) and field capacity be better for a sandy soil or a fine-textured soil with a well-developed structure?

11. If a porous sample 5 cm in length is placed in a centrifuge to determine $p_c(S)$, what length of sample would have to be employed by the "long-column" method to obtain the same range of p_c? Assume the centrifuge is operated at the usual angular velocity.

12. Would you expect the long-column procedure for determining $p_c(S)$ to be popular in petroleum industry for characterizing porous rocks? Explain.

13. Would you expect a piezometer to be equally as effective as a tensiometer for measuring the pressure of water in soil within the root zone of crops? Explain.

14. Why would one want to determine the pressure of soil water within the root zone of crops?

15. Would you expect the value of λ in the Brooks-Corey equation to be larger for a medium consisting of freshly crushed quartz particles, or for sand taken from a swift flowing river bed? Assume that both media have the same grain-size distribution.

16. Other things being equal, would you expect a packing technique that results in a smaller porosity to result in a larger or smaller value of λ for a given granular sedimentary material in a laboratory column.

17. If the value of $p_c/\rho_w g$ measured with air into water is 50 cm for a particular sandstone, what would you expect p_e (expressed in atmospheres) to be for mercury into air?

18. Consider a soil containing static water, with a water table at a fixed elevation beneath the surface. If the soil texture varies in horizontal planes, describe qualitatively how you would expect the saturation to vary in horizontal planes in respect to the value of p_e.

19. Explain why the term "pseudo" equilibrium is used to describe the condition of water in fallow soils.

20. It is desired to locate the lower edge of the oil body and also the lower edge of a gas body in a static petroleum reservoir. Assuming that it is possible to measure the

pressure of brine, oil and gas independently at the bottom of an observation well, explain where such a well might be drilled to obtain the necessary data from a single hole. Describe all the data that would be needed and explain how it would be used to make the necessary calculations.

21. The Brooks methods of determining $p_c(S)$ on the wetting cycle reduces the importance of excellent contact between sample and capillary barrier as compared to the pressure-cell procedure. Explain.

22. When using a centrifuge to obtain the moisture equivalent of a sample of porous rock, it is very important to obtain good contact between the sample and a porous plug at the outer face of the sample. Explain.

23. Sketch a hypothetical relationship $p_c(S)$ such that S_y would not be a function of depth to the water table. What would be the value of p_d for this case?

24. The U.S. Bureau of Reclamation assumes a constant value of S_y in their procedure for the design of drainage systems. List all of the conditions that are necessary for this to be a valid approximation.

25. Would you expect that the use of mercury injection (to obtain the $p_c(S)$ function for characterizing the pore-size distribution) to be more valid for a sandstone containing negligible clay or for one that contained considerable montmorillonite? Explain.

CHAPTER III

EQUATIONS OF FLUID FLUX IN POROUS MEDIA

3.1 FLUID MOTION

According to Newton's second law, the rate of change of momentum with respect to time (of an element of mass) is equal to the resultant force acting on the element. If the mass of the element does not change with time, Newton's law implies that the product of mass and acceleration is equal to the force. In the following analyses, Newton's law is written in respect to a fluid particle, as defined in Section 1.5.2. It is assumed that the particle is undergoing negligible divergence.

In this case, the only force acting on the particle, other than the driving forces that act also on static elements, is fluid shear. With this simplification, Newton's law indicates that, for any direction i,

$$\rho \frac{du_i}{dt} = \rho g_i - \frac{\partial p}{\partial x_i} + F_i(\text{shear}) \quad . \tag{3.1}$$

In Equation 3.1, all terms are expressed on a <u>per volume</u> basis. The left side represents the product of mass and acceleration, and the right side is a summation of force components in the i direction.

The first two terms on the right represent the driving forces as described in Section 2.1.1. The terms representing acceleration and shear force components, however, require further explanation.

3.1.1 Fluid velocity

- The component of velocity u_i refers to motion of the center of mass of a specified volume (the fluid particle). It does not refer to the motion of individual molecules or ions, or to any species of molecules that constitute the fluid mass. It is required, for the analysis presented here, that the element contains the same mass at all times as it moves in space, but the individual molecules contained in the element are continuously interchanging with those of neighboring elements. Consequently, the fluid particle is an abstraction rather than a physical entity.

It is assumed that the derivatives of u_i (of any order) with respect to the space coordinates exist. Thus, u_i is regarded as an analytic function of the space coordinates throughout a fluid phase <u>including the solid boundaries</u>. This assumption has limitations as explained in Section 3.7.2.

3.1.2 Fluid acceleration - The derivative du_i/dt is the "total" component of acceleration in the direction i. Noting that, in general, u_i is a function of orthogonal space coordinates, x_i, x_j, and x_k, as well as t,

$$du_i = \frac{\partial u_i}{\partial x_i} dx_i + \frac{\partial u_i}{\partial x_j} dx_j + \frac{\partial u_i}{\partial x_k} dx_k + \frac{\partial u_i}{\partial t} dt \quad .$$

Dividing by dt gives

$$\frac{du_i}{dt} = (u_i \frac{\partial u_i}{\partial x_i} + u_j \frac{\partial u_i}{\partial x_j} + u_k \frac{\partial u_i}{\partial x_k}) + \frac{\partial u_i}{\partial t}$$

in which u_i, u_j and u_k are velocity components in the x_i, x_j and x_k directions respectively.

Writing this with the <u>summation convention</u> results in

$$\frac{du_i}{dt} = u_j \frac{\partial u_i}{\partial x_j} + \frac{\partial u_i}{\partial t} \quad , \qquad 3.2$$

the repeated subscript j indicating a summation over three orthogonal coordinate directions, i, j, and k.

The first term on the right of Equation 3.2 is the <u>convective acceleration</u> which is due to velocity variations (direction as well as magnitude) with respect to position in space. The term $\partial u_i/\partial t$ refers to the variation of u_i (at a particular point in space) with respect to time. It is called <u>local acceleration</u>.

3.1.3 Fluid shear - A resisting force which acts tangential to the surface of moving particles is called <u>shear</u>. It is proportional to the area over which it acts, and depends on the component of velocity gradient normal to the plane in which it acts. Shear on a particular face of a fluid element (say an element consisting of a cube) is a force in the direction of motion of a faster moving adjacent element. If the motion of the adjacent element is slower, the force of shear is in the opposite direction on the face under consideration. The force/area is called <u>intensity of shear</u> τ.

Shear is a tensor quantity. In order to specify τ at a particular point, it is necessary to identify the orientation of the plane under consideration and the direction within this plane as well as its magnitude. Thus, τ_{ij} means the intensity of shear (at a particular point) in a plane normal to i in the direction j.

In analyzing the resultant force on a volume element, say a cube, it is necessary to consider only the shear at the faces of the cube. This is because shear on other planes within the cube exerts equal force in opposite directions on adjacent parts of the same element; that is, the resultant effect of shear on all but the surface faces is zero in respect to a particular element under consideration.

The resultant of shear on a fluid particle of any shape can be expressed generally by

$$\underline{F} = \int_{A_p} \underline{\sigma} \times \underline{da} \qquad 3.3$$

in which $\underline{\sigma}$ is the surface force, $\underline{da}$ is a vector having the direction of the outer normal to the surface and the magnitude of a differential segment of the surface, and A_p is the entire surface area of the particle. Note that $\underline{\sigma}$ is not τ, because it may have a normal as well as a tangential component. The cross product $\underline{\sigma} \times \underline{da}$, however, evaluates only the tangential components. Unlike $\underline{\sigma}_c$ appearing in Equation 2.1, $\underline{\sigma}$ is not conservative, and cannot be related to volume and temperature in an equation of state. Furthermore $\underline{F}$, unlike p as defined by Equation 2.1, is a vector.

Although, Equation 3.3 is useful for visualizing the effect of shear in creating a resultant force on a volume element, it does not help in evaluating this force. A more useful expression for F_i, the component of shear in the i direction, can be developed in respect to the spatial distribution of velocity. A derivation of the latter expression, when a velocity component is varying in three coordinate directions, is beyond the scope of this text. Such a derivation is given in all advanced texts on fluid mechanics. However, a derivation of F_i, for the simple case of 1-dimensional flow, provides a relationship in the same form as the general relationship and may provide some insight into the way shear acts.

A case is considered in which u_i is varying in an orthogonal direction j and components u_j and u_k are zero. The space coordinate in the j direction is x_j. In this case it is assumed that

$$\tau_{ji} = \mu \frac{\partial u_i}{\partial x_j} ,$$

in which μ is the viscosity of the fluid. The derivative $\partial u_i/\partial x_j$ represents the rate of angular deformation (in planes normal to k) of volume elements of the fluid. The assumption (that this is linearly related to the shear stress) means that the fluid is a Newtonian viscous fluid. For such a fluid, in the general case of 3-dimensional flow, it is possible to show that

$$\tau_{ji} = \mu\left(\frac{\partial u_j}{\partial x_i} + \frac{\partial u_i}{\partial x_j}\right) .$$

However, in the case under consideration, the first term in the parenthesis is zero because u_j is zero.

A particle having the shape of a cube with a volume $(\delta\ell)^3$ is considered, as illustrated in Figure 3-1.

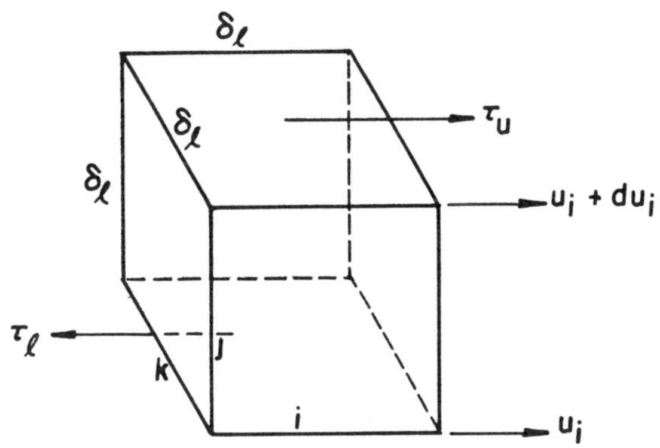

Figure 3-1. Shear on a cube undergoing 1-dimensional deformation.

The direction of shear in the upper face τ_u is indicated in Figure 3-1, as is the direction of shear in the lower face τ_ℓ. These directions imply that du_i, also shown in the figure, is positive. In other words, $\partial u_i/\partial x_j$ is positive for the case illustrated.

For this case,

$$F_i = [(\tau_{ji})_u - (\tau_{ji})_\ell] \frac{(\delta\ell)^2}{(\delta\ell)^3} .$$

Since

$$\tau_{ji} = \mu \frac{\partial u_i}{\partial x_j} ,$$

the term in parenthesis becomes

$$\mu \left[\frac{\partial}{\partial x_j}\left(\frac{\partial u_i}{\partial x_j}\right)\right]\delta \ell$$

and

$$F_i = \mu \frac{\partial^2 u_i}{\partial x_j \partial x_j} \quad . \qquad 3.4$$

Although Equation 3.4 was derived for a case in which all velocity components except u_i are zero, and u_i is varying only in the j direction, it has the same form as the expression for the case of 3-dimensional flow. In the latter case, however,

$$\frac{\partial^2 u_i}{\partial x_j \partial x_j} = \frac{\partial^2 u_i}{\partial x_i^2} + \frac{\partial^2 u_i}{\partial x_j^2} + \frac{\partial^2 u_i}{\partial x_k^2}$$

which conforms with the summation convention.

In vector notation,

$$\frac{\partial^2 u_i}{\partial x_j \partial x_j} \equiv \nabla^2 u_i \quad .$$

3.1.4 Equation of fluid motion - Substituting Equation 3.2 and 3.4 into Equation 3.1 gives

$$\rho\left(u_j \frac{\partial u_i}{\partial x_j} + \frac{\partial u_i}{\partial t}\right) = \rho g_i - \frac{\partial p}{\partial x_i} + \mu \frac{\partial^2 u_i}{\partial x_j \partial x_j} \quad . \qquad 3.5$$

Equation 3.5 is a special case of the Navier-Stokes equation of fluid motion, that is, the case for Newtonian viscous fluids undergoing negligible divergence.

This equation usually can be simplified further for application to flow in porous media. If the medium is homogeneous, the convective acceleration term

$$u_j \frac{\partial u_i}{\partial x_j} \quad ,$$

when integrated over a macroscopic volume, is zero for uniform rectilinear macroscopic flow, [Hubbert (1940)]. That is, the velocity (statistically) is unchanged in respect to both magnitude and direction as a result of fluid passing through a macroscopic volume element of the medium. If u_i is small, as is the usual case, $\partial u_i/\partial t$ is also small. It usually is assumed that both local and convective terms are small even for non-homogeneous porous media and for flow that is not uniform or rectilinear.

Based upon that assumption, the equation for motion of fluids in porous media becomes

$$\frac{\partial p}{\partial x_i} - \rho g_i = \mu \frac{\partial^2 u_i}{\partial x_j \partial x_j} . \qquad 3.6$$

It is emphasized that both Equation 3.5 and 3.6 are written in respect to fluid particles which are totally within one fluid phase or another. Elements that include some of each phase are excluded.

3.2 CONCEPT OF A POTENTIAL

It may sometimes be convenient to express the force component acting on a fluid particle as the negative gradient of a scalar quantity called a force potential. The potential is related to the concept of potential energy, implying a capacity to do work. However, it is possible to define a force potential only in respect to force fields that are conservative, [Hubbert (1940)].

The potential energy of an element of fluid (with respect to a particular field of force) is a function of its position in the field. The value of the potential (when it exists) is a measure of the work that would be done by a particular force field on a fluid particle in moving from a given position to a datum position, providing this work could be done reversibly.

The existence of a potential of a particular force field does not depend, however, upon whether or not a fluid particle physically could be moved from one point to another reversibly. The existence depends, instead, upon the way the force is related to the space coordinates. A force field for which a scalar potential can be defined is said to be conservative.

A necessary and sufficient condition for a force field to be conservative is that the work integral

$$\oint \underline{f} \cdot \underline{ds} = 0$$

where $\underline{f}$ is the force considered, and $\underline{ds}$ is a differential displacement vector in the field. The circle indicates that the integral has been taken about any closed path. If this condition is satisfied, it is possible to define a scalar Φ by

$$\Phi \equiv \int_{\underline{s}_o}^{\underline{s}} - \underline{f} \cdot \underline{ds} \qquad 3.7$$

where Φ is the potential at the position $\underline{s}$, and $\underline{s}_o$ represents the position of a datum in respect to which Φ is evaluated. The negative sign of the integrand conforms with the traditional sign convention for potentials. Clearly, the definition is ambiguous if the work integral around every closed path is not zero.

3.2.1 Pressure potential - The significance of a potential (when its definition is valid) is that the negative gradient of the potential represents the force acting at any point in the system. Forces that are a consequence of motion (in this case shear) and result in the dissipation of energy in the form of heat are non-conservative. Driving forces (those that tend to produce the motion) are sometimes conservative. It is clear, for example, that

$$p_{\underline{s}} \equiv \int_{\underline{s}_o}^{\underline{s}} - \left(- \frac{\partial p}{\partial \underline{s}}\right) \underline{ds} \quad ,$$

if $p_{\underline{s}_o}$ is set equal to zero.

Pressure, therefore, is a potential having the dimensions of energy/volume, the negative gradient of which is the force resulting from the spatial distribution of pressure. Note that whereas p is a scalar, the gradient of p,

$$\frac{\partial p}{\partial x_i} \underline{e}_i \quad ,$$

is a vector and as such has a particular direction associated with it, depending upon the relative magnitude of its components,

$$\frac{\partial p}{\partial x_i} \quad .$$

In physics, it is customary to define force potentials having the dimensions of energy/mass only. It is informative, therefore, to define a pressure potential by

$$\Phi_p \equiv \int_{s_o}^{s} - [- \frac{\partial(p/\rho)}{\partial \underline{s}}] \, \underline{ds} \quad .$$

Note, however, that the definition is valid only for a case in which ρ is constant with respect to $\underline{s}$, or varies only with p. The latter is called the barotropic case.

The error resulting from attempting to regard p/ρ as a force potential (for a case of ρ varying independently with the space coordinates) can be understood by arbitrarily defining Φ_p as p/ρ. Note that

$$- \frac{\partial(p/\rho)}{\partial x_i} = - \frac{1}{\rho} \frac{\partial p}{\partial x_i} - p \frac{\partial(1/\rho)}{\partial x_i} \quad .$$

Clearly, the component of force in the i direction (due to the pressure gradient) is $-(1/\rho)(\partial p/\partial x_i)$, but the second term on the right is not a force component, and may be regarded as an error term resulting from arbitrarily defining Φ_p as p/ρ.

3.2.2 Gravitational potential - Similarly, the potential Φ_g due to gravity is given by

$$\Phi_g \equiv \int_{\underline{s}_o}^{s} - \underline{g} \cdot \underline{ds}$$

where $\underline{g}$ is the force/mass due to gravity. Where the elevation difference is sufficiently small that changes in $\underline{g}$ can be ignored,

$$\Phi_g = gh \quad ,$$

h being the elevation difference between $\underline{s}$ and $\underline{s}_o$. Note that here Φ_g has the dimensions of energy/mass.

It is informative to attempt a definition of a gravity potential Φ'_g, having the dimensions of energy/volume, by

$$\Phi'_g \equiv \rho g h \quad .$$

Taking the negative derivative of Φ'_g with respect to x_i gives

$$- \frac{\partial \Phi'_g}{\partial x_i} = - \rho g \frac{\partial h}{\partial x_i} - gh \frac{\partial \rho}{\partial x_i} \quad ,$$

providing g is regarded as a constant. Clearly, the first term on the right, is a valid expression for the component of force/volume due to gravity, but the second term is, just as clearly, an error term having no relation to a force component.

The magnitude of $-gh(\partial\rho/\partial x_i)$ depends upon the elevation h as well as upon the component of the density gradient. Its importance depends on its magnitude relative to that of $-\rho g(\partial h/\partial x_i)$ and $-(\partial p/\partial x_i)$. A potential term evaluated as ρgh, therefore, should be used with caution, and in particular, it should not be assumed that the error is unimportant when the variation in density is small. The error, if the system is large, can be important even with small density differences. It also can be important when $\partial h/\partial x_i$ and $\partial p/\partial x_i$ are both of the same order as $\partial\rho/\partial x_i$.

3.2.3 Combined pressure and gravity potential - When it is possible to define two or more force potentials in respect to their individual force fields, it is permissible to add the potentials to obtain a combined potential Φ'. In this case the negative of $\partial\Phi'/\partial x_i$ gives the combined force component in the direction i.

It is not possible to combine Φ_p and Φ_g regardless of the dimensions in which these potentials are expressed, unless ρ is constant or depends only on p. This is because either Φ_p or Φ_g is not defined in the general case. For the special case of a constant density, however, it may be very convenient to combine Φ_p and Φ_g. Clearly, to do this both Φ_p and Φ_g must have the same dimensions and refer to the same reference element. Two ways of doing this are in common use:

(1) A potential p^* is defined by

$$p^* \equiv p + \rho gh \qquad 3.8$$

which has the dimensions of energy/volume and is called <u>piezometric pressure</u>.

(2) A potential H is defined by

$$H \equiv \frac{p}{\rho g} + h \qquad 3.9$$

which has the dimensions of energy/weight (that is, length), and is called <u>piezometric head</u>.

Occasionally some authors have used a third combined potential,

$$\Phi' \equiv \frac{p}{\rho} + gh \quad , \qquad 3.10$$

having the dimensions of energy/mass, but the latter is seldom used in either hydraulics or fluid mechanics.

The question of whether p^* or H is the more convenient depends upon the particular problem being analyzed, and each is used in this text where appropriate.

Many of the early investigators, for example Richards (1928), called P^*(or H) a "total" potential. This is a valid concept provided ρ is constant. In later years, however, some investigators [notably Edlefsen and Anderson (1943)] attempted to extend the concept of "total potential" to include terms not pertaining to forces on fluid particles as defined in this text. For example, they included terms pertaining to the motion (of individual species of molecules, e.g., water) in response to osmotic and temperature effects. As pointed out by Corey and Kemper (1961), this procedure led to erroneous conclusions that have been widely disseminated among soil scientists. The concept of a "total" potential is valid only if:

(1) All force fields for which potentials are defined are conservative.
(2) All potentials which are combined are in respect to a common reference element. It is not valid, for example, to add an osmotic potential (in respect to water "molecules") to a potential relating to fluid "particles".
(3) All potentials are force potentials in the Newtonian sense. For example, so-called "forces" pertaining to molecular diffusion in response to temperature and concentration gradients have no independent effect on the acceleration of the center of mass of a fluid particle.

It is better to regard convective and diffusion processes as separate transport mechanisms and to combine the results, after each has been evaluated independently [A. T. Corey and A. Klute (1985)].

3.3 FLOW IN MODELS OF A POROUS MEDIUM

Writing Equation 3.6 in terms of the potential function p^* results in

$$\frac{\partial p^*}{\partial x_i} = \mu \frac{\partial^2 u_i}{\partial x_j \partial x_j} \quad , \qquad 3.11$$

which is a valid approximation for flow in porous media under ordinary potential gradients provided the fluid density is constant.

It would be desirable to use this equation to describe flow through the pore space of porous media. However, to do this, it is necessary to describe the geometric boundaries of the pore space in such a way that the differential equation can be solved. The pore space, however, has a very complex geometry. Even if the complex geometry could be described, it is doubtful that the equation could be solved for such boundaries.

The approach used in the following analyses is to select highly idealized and simplified models that, never-the-less, can be classified as porous media, and to solve Equation 3.11 for these cases. By this procedure, it is hoped that an insight is provided in respect to the probable behavior of fluids in more complex porous media.

In addition to the assumptions, previously discussed, which were accepted for the derivation of Equation 3.11, additional assumptions are made in describing the boundary conditions for the flow. These are:

(1) At boundaries where a fluid is in contact with a solid, the fluid velocity relative to the boundary is zero. If this were not true, it would be necessary to postulate that the derivative of $\underline{u}$ with respect to the normal to the boundary is infinite. This is not in conformance with the continuum assumption that $\underline{u}$ is an analytic function of the space coordinates at all points including the boundaries. Also, it would imply an infinite shear at the boundary.
(2) At boundaries where a liquid is in contact with a gas, the shear is assumed to be negligible. This is based on the fact that the viscosity of gases are less than that of common liquids by about three orders of magnitude.
(3) Symmetry of velocity distribution is assumed where there is no reason to postulate a lack of symmetry.

3.3.1 Flow in a film - The first model considered is a liquid flowing steadily over a flat solid boundary in a film of uniform thickness d. The flow is 1-dimensional in the direction i; that is, u_j and u_k are zero. Such a model is illustrated in Figure 3-2, in which a section of the film (in a plane parallel to i and normal to the solid boundary) is shown.

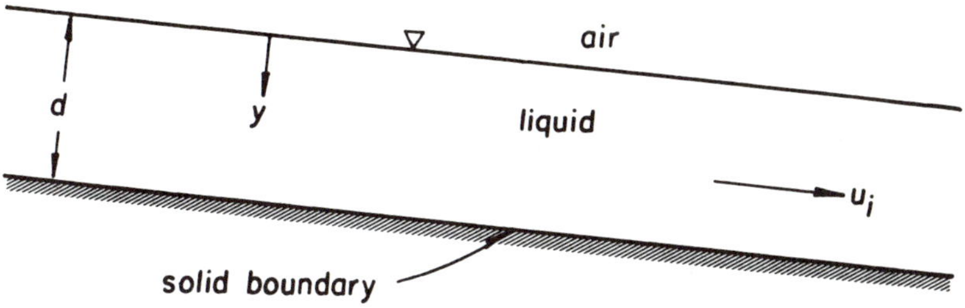

Figure 3-2. Flow in a film over a flat solid boundary.

Because all components of $\underline{u}$ orthogonal to u_i are zero, the right side of Equation 3.11 can be written as an ordinary second derivative, that is

$$\mu \frac{d}{dy} \left(\frac{du}{dy}\right) ,$$

in which y is a coordinate measured from the surface of the film in a direction normal to the solid boundary and also to the direction of u. The subscript i is dropped because $\underline{u}$ has zero components in orthogonal directions. Likewise, the left side of Equation 3.11 can be written as an ordinary derivative. The rationale for this is that since there is no flow orthogonal to i, there is no component of the potential gradient orthogonal to i. Letting x be the coordinate in the i direction results in

$$\frac{dp^*}{dx} = \mu \frac{d}{dy} \left(\frac{du}{dy}\right) .$$

Also, the derivative dp^*/dx does not vary with y. If dp^*/dx did vary with y, this would imply that p^* varies with y which is not the case. Integrating with respect to y gives

$$\frac{dp^*}{dx} y = \mu \frac{du}{dy} + c .$$

Since τ at an air boundary is considered to be zero

$$\frac{du}{dy} = 0 \quad \text{at} \quad y = 0 ,$$

therefore

$$\frac{y}{\mu} \frac{dp^*}{dx} = \frac{du}{dy} .$$

Integrating a second time gives

$$\frac{y^2}{2\mu}\frac{dp^*}{dx} = u + c' \quad .$$

Evaluating c' on the basis that u is zero at the solid boundary (where $y = d$) gives

$$u = -\frac{1}{2\mu}\frac{dp^*}{dx}(d^2 - y^2) \quad , \qquad 3.12$$

which indicates that the velocity distribution is parabolic.

The average velocity $\bar{u}$ is defined by the expression

$$\bar{u} \equiv \frac{1}{d}\int_o^d u\, dy \quad .$$

Substituting Equation 3.12 into this expression and integrating gives

$$\bar{u} = -\frac{d^2}{3\mu}\frac{dp^*}{dx} \quad . \qquad 3.13$$

The discharge through the film per unit width normal to the flow is given by $\bar{u}\,d$.

3.3.2 Flow through slits - The second case considered also involves a 1-dimensional flow situation. The model is like that for the film case except that the upper flow boundary, as well as the lower, is solid. This situation is illustrated in Figure 3-3. Again a section normal to the boundaries and parallel to the flow is considered. The coordinate in the direction of flow is designated as x, and the coordinate orthogonal to both x and the boundaries is y. The origin for y is at one of the solid boundaries.

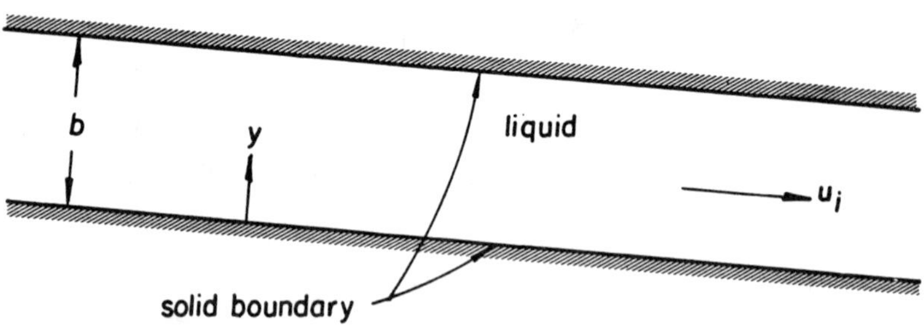

Figure 3-3. Flow through a slit of uniform thickness.

For this case, the flow equation

$$\frac{dp^*}{dx} = \mu \frac{d}{dy} \left(\frac{du}{dy}\right) ,$$

again reduces to an ordinary differential equation and dp^*/dx is a constant with respect to both x and y. Integrating with respect to y twice and using the assumption, that because of symmetry, du/dy is zero where y is $b/2$, and that u is zero where y is either zero or b, the result is

$$u = \frac{1}{2\mu} (y^2 - by) \frac{dp^*}{dx} . \qquad 3.14$$

Also,

$$\bar{u} = - \frac{b^2}{12\mu} \frac{dp^*}{dx} . \qquad 3.15$$

3.3.3 Flow through tubes with circular cross-sections

In the case of flow through circular tubes, Equation 3.11 cannot be reduced to an ordinary differential equation in an orthogonal coordinate system. However, in cylindricial coordinates it is an ordinary equation, assuming that the flow is symmetrical about the axis of the tube. Such a tube is shown in Figure 3-4.

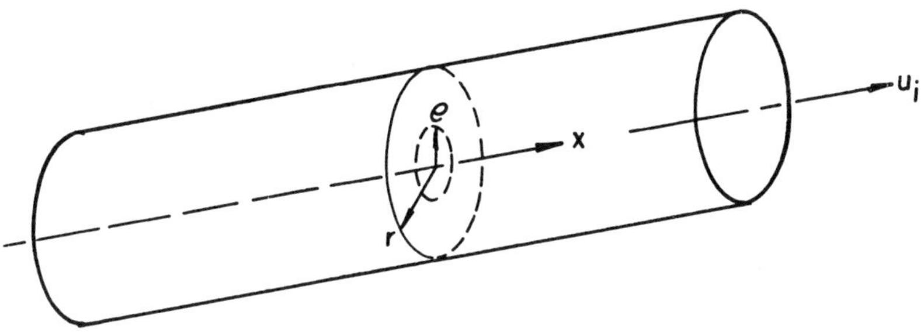

Figure 3-4. Flow through a capillary tube with circular cross section.

In cylindrical coordinates, the operator

$$\frac{\partial^2}{\partial x_j \partial x_j}$$

becomes

$$\frac{1}{\rho} \frac{\partial}{\partial \rho} \left(\rho \frac{\partial}{\partial \rho}\right) + \frac{1}{\rho^2} \frac{\partial^2}{\partial \theta^2} + \frac{\partial^2}{\partial x^2} ,$$

in which ρ is the radial coordinate with origin at the axis. For the symmetrical case, the two terms on the right are zero. Consequently,

$$\frac{dp^*}{dx} = \mu[\frac{1}{\rho}\frac{d}{d\rho}(\rho\frac{du}{d\rho})] \quad .$$

Again using the argument that $du/d\rho$ is zero at the center and u is zero at the solid boundary, the result is

$$\bar{u} = -\frac{r^2}{8\mu}\frac{dp^*}{dx} \quad , \qquad 3.16$$

r being the radius of the tube. Equation 3.16 is known as Poiseuille's equation.

3.3.4 Generalized equation for flow through straight conduits

- By induction, after examining Equations 3.13, 3.15 and 3.16, it is possible to conclude that an equation of the form

$$\bar{u} = -\frac{D^2}{k_s\mu}\frac{dp^*}{dx} \qquad 3.17$$

describes $\bar{u}$ for small uniform conduits in general. In Equation 3.17, D is a characteristic length dimension describing the size of cross-section and k_s is a factor depending upon the shape and the dimension selected. For example k_s would be 32 instead of 8 if the diameter had been used instead of the radius to describe the size of tube shown in Figure 3-4.

Equation 3.17 states that the average velocity (and, consequently, the discharge) is proportional to the driving force in a small straight tube. It also states that the constant of proportionality is directly related to a characteristic length squared and is inversely related to fluid viscosity.

3.4 HYDRAULIC RADIUS

Since the factor k_s in Equation 3.17 depends upon the choice of a length dimension to describe the tube size, it is desirable to select a dimension that can be defined in an unambiguous way for all cross-sectional shapes. It is also desirable to select a dimension that correlates well with discharge for a given driving force and so minimizes the range of k_s.

The length used for this is well known in hydraulic engineering as the hydraulic radius which is defined by

$$R \equiv \frac{A}{wp} \qquad 3.18$$

in which A is the area of cross-section normal to flow and wp is the length of the wetted perimeter of the flow section in the plane of A.

In respect to the tubes analyzed in Sections 3.3.1, 3.3.2 and 3.3.3, the value of R is given by d, b/2 and r/2 respectively. Replacing the dimensions d, b and r by their equivalent in terms of R, the equations for $\bar{u}$ for the film and slit become

$$\bar{u} = -\frac{R^2}{3\mu}\frac{dp^*}{dx} , \qquad 3.19$$

and that for the circular tube becomes

$$\bar{u} = -\frac{R^2}{2\mu}\frac{dp^*}{dx} . \qquad 3.20$$

For these particular cases, therefore, the range of k_s has been reduced by the use of R as a characteristic length to a factor that varies from 2 to 3.

The use of R does not reduce the range of k_s for all shapes, however. To visualize this, consider a tube of about the same cross-sectional area as that shown in Figure 3-4, but with a shape as illustrated in Figure 3-5.

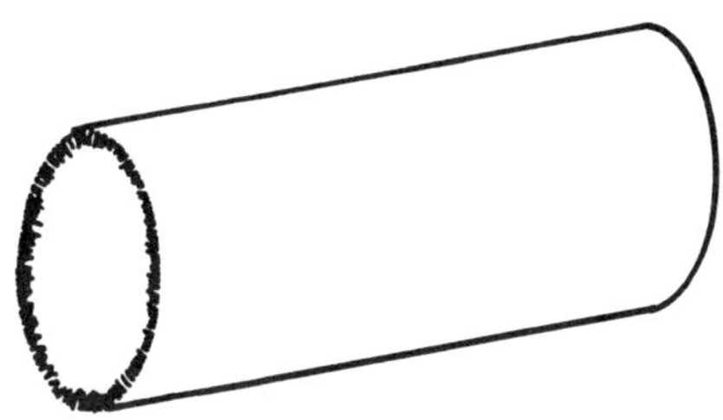

Figure 3-5. Cross-section of tube with extended wetted perimeter.

For a case of laminar flow of a viscous fluid, it is intuitively clear that $\bar{u}$ is not greatly different from that for a tube with a smooth bore of the same cross-sectional area. As long as turbulence is not a factor, the spatial distribution of velocity in the main part of the tube is affected only slightly by the local variations near the perimeter. Yet the value of R may be smaller by a large factor in the tube with the extended perimeter. It is clear that the concept of R must be used with caution.

3.4.1 Effect of tube-size distribution - One way to overcome the problem posed by shapes analogous to that shown in Figure 3-5 is to regard the flow region in the vicinity of the rough perimeter as being separate from the main portion of the tube. It would be theoretically possible to evaluate an R for each little crevice around the boundary and add their individual flow contributions, which in this case would be negligible. An analogous procedure is used in hydraulics in evaluating flow through a river channel of irregular cross-sectional shape.

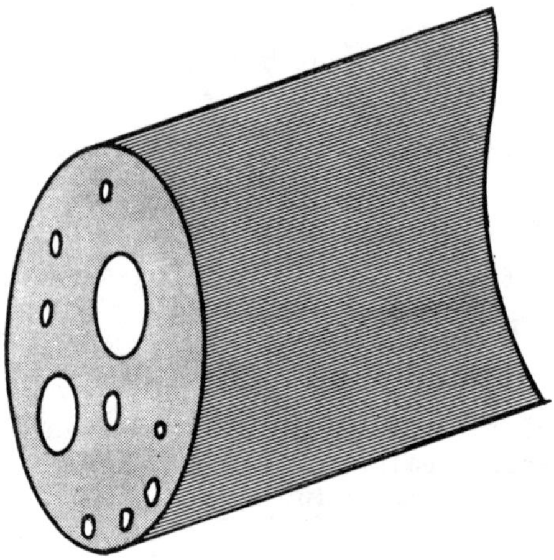

Figure 3-6. Model of porous medium consisting of a cylinder of solid material with a number of round holes bored parallel to axis of cylinder.

In a similar way, it is possible to evaluate the flow through a model consisting of straight circular tubes of varying diameter as shown in Figure 3-6. Assuming that Equation 3.20 applies to each individual tube, it is possible to find a valid value for $\bar{u}$ for the model as a whole by adding the contribution to the discharge of each individual tube and then dividing by the total cross-sectional area. Applying Equation 3.20 to flow through each hole in this model,

$$Q_i = - \frac{4\pi R_i^4}{2\mu} \frac{dp^*}{dx} ,$$

where Q_i is the discharge through a particular hole with hydraulic radius R_i. Consequently, for the bundle as a whole,

$$Q = \sum_{i=1}^{n} Q_i = -\frac{4\pi}{2\mu}\frac{dp^*}{dx}\sum_{i=1}^{n} R_i^4 .$$

Using the relationship that the total cross-sectional area of the bulk system is equal to the total pore area divided by ϕ gives

$$q = -\frac{\phi\overline{R^2}}{2\mu}\frac{dp^*}{dx}$$

in which q is the ratio Q/A, A being the total cross-sectional area of the cylinder, and

$$\overline{R^2} \equiv \sum_{i=1}^{n} R_i^4 \Big/ \sum_{i=1}^{n} R_i^2 . \qquad 3.21$$

Equation 3.21 defines a quantity $\overline{R^2}$ having the physical significance of a "weighted mean" value of R^2.

It is important to observe that $\overline{R^2}$ is, in general, larger than the quantity $\bar{R}^2$ where $\bar{R}$ is a mean value of R given by the ratio of total tube volume to the total internal surface area of tube. This situation can be verified by considering a model with two holes with R_i equal to 1 and 3 respectively. For this case, $\bar{R}^2$ is 6.25 where $\overline{R^2}$ is 8.2. The disparity becomes much greater as the ratio of larger to smaller R_i values increases. The disparity is enormous when the R_i differ by orders of magnitude.

3.4.2 Hydraulic radius related to pore-size - As explained in Section 3.4.1, an expression which gives an average value of R, for a tube of non-uniform cross-section, is the internal volume of the tube divided by its internal surface area. A statistically equivalent expression applicable to a porous medium is ϕ/s.

If the volume of the reference element becomes as small as a portion of the pore space between individual grains, the volume divided by the internal surface approaches the value of R for that portion of the pore space. As explained in Section 2.4, one way to evaluate A/wp for a portion of pore space is to assume that it is inversely related to the value of p_c when an interface is stretched across the section of pore space under consideration.

The capillary pressure, therefore, should be a measure of the largest pore-size or R value for that part of the pore space occupied by the wetting fluid at a particular saturation. It is also a measure of the smallest R value of that part of the pore space occupied by the non-wetting fluid.

3.4.3 The value of R^2 as a function of S

- It should be intuitively clear that as S decreases, R_{max} for the wetting phase (or R_{min} for the non-wetting phase) also decreases. In respect to size, the pore space is characterized by $\overline{R^2}$ for all of the space occupied by a particular fluid.

For the wetting phase,

$$\overline{R_w^2} = \frac{\sigma^2 \cos^2\alpha}{p_d^2\, S} \int_o^S \left(\frac{p_d}{p_c}\right)^2 dS \quad , \qquad 3.22$$

and for the non-wetting phase,

$$\overline{R_{nw}^2} = \frac{\sigma^2 \cos^2\alpha}{p_d^2(1-S)} \int_S^{1.0} \left(\frac{p_d}{p_c}\right)^2 dS \quad . \qquad 3.23$$

Equations 3.22 and 3.23, of course, must be modified for a 3-phase fluid system. Equations similar to Equations 3.22 and 3.23 were first proposed by Purcell (1949). Purcell based his derivation upon a model of a porous medium consisting of a bundle of capillary tubes.

3.5 TORTUOSITY

A porous medium is not, however, a bundle of straight tubes, nor even a bundle of sinuous tubes. An appropriate model must consist of a network of interconnected channels. Fluid particles flowing through granular porous media, however, do follow a sinuous path. Although in granular media the pore space is interconnected, the effect of the sinuous path (at least, some aspects of it) can be deduced by considering a model consisting of a single sinuous channel, as shown in Figure 3-7.

It is possible to measure the length ℓ in the direction i from the point where the pressure is p_1 to a point where it is p_2. However, it is not possible to determine directly the distance ℓ that fluid particles must move in passing

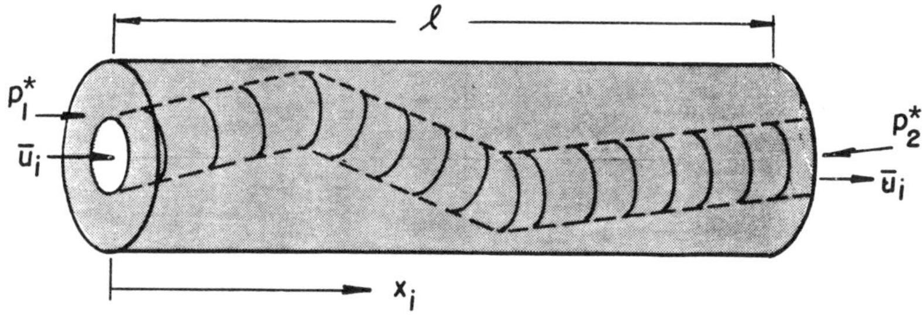

Figure 3-7. Model consisting of a solid with a single sinuous channel.

from point 1 to point 2. The path length is designated by ℓ_e, implying that this is the effective distance of flow between point 1 and point 2. The fact that ℓ_e/ℓ is greater than 1.0 has two important consequences:

(1) The average value of $-\partial p^*/\partial \ell_e$ acting on the fluid is

$$\frac{p_1^* - p_2^*}{\ell_e}$$

or

$$\frac{\ell}{\ell_e}\left(\frac{p_1^* - p_2^*}{\ell}\right) .$$

(2) The average component of velocity in the i direction is

$$\bar{u}_i = \frac{\ell}{\ell_e}\bar{u}_t ,$$

in which $\bar{u}_t$ is the average velocity tangential to the flow path.

To calculate $\bar{u}_i$ (knowing only R for the tube and the values of p_1 and p_2), it is possible to write

$$\bar{u}_t = \frac{R^2}{k_s\mu}\left(\frac{p_1^* - p_2^*}{\ell_e}\right)$$

or

$$\bar{u}_i \left(\frac{\ell_e}{\ell}\right) = \frac{R^2}{k_s \mu} \frac{\ell}{\ell_e} \left(\frac{p_1^* - p_2^*}{\ell}\right) .$$

The latter equation is frequently written in differential form as

$$\bar{u}_i = - \frac{R^2}{k_s T \mu} \frac{\partial p^*}{\partial x_i} , \qquad 3.24$$

in which T is called tortuosity $(\ell_e/\ell)^2$. The ratio ℓ/ℓ_e may also be interpreted as the average cosine of the angle of $\bar{u}_t$ with respect to the direction i.

According to Wyllie and Spangler (1952), T for typical granular porous media when fully saturated with a single fluid is about 2, which implies that the average angle of the fluid path in respect to the macroscopic flow direction is about 45°. This also was observed to be a good approximation by Sullivan and Hertel (1942). By assuming a "random walk" model with the preferred angle being zero, and the probability decreasing to zero at an angle of π, it is possible to deduce that the average angle is $\pi/4$ which agrees with the observation that the value of T is about 2 for the fully saturated case. Actual values of T may differ from 2 significantly, however, for non-isotropic media.

3.5.1 Tortuosity as a function of saturation - As the wetting-phase saturation of a porous medium decreases, fluid particles of the wetting phase must take an increasingly longer path in moving between two points. This is because the particles cannot take a direct route across pore spaces because the central portion of the spaces are occupied by the non-wetting phase. After sufficient desaturation takes place, the non-wetting phase has a finite tortuosity, and this tortuosity decreases with decreasing liquid saturation.

It has been determined experimentally by Burdine (1952) and Corey (1954) that the tortuosity of a wetting phase T_w is inversely related to S_e^2; that is,

$$\left(\frac{T_{1.0}}{T_{S_e}}\right)_w \cong S_e^2 . \qquad 3.25$$

in which $T_{1.0}$ is the tortuosity of the wetting phase when S_e is 1.0. The same investigators also found that for the non-wetting phase

$$\left(\frac{T_{1.0}}{T_{S_e}}\right)_{nw} \cong (1-S_e)^2 \quad . \qquad 3.26$$

Both Equation 3.25 and 3.26, however, are valid only for isotropic media. Wyllie and Gardner (1958) later deduced the same $T(S_e)$ functions from probability considerations using a capillaric model for a porous medium consisting of randomly interconnected bundles of tubes.

3.6 KOZENY-CARMAN EQUATION

Kozeny (1927) and later Carman (1937) and Fair and Hatch (1933) developed an equation for $\bar{u}_i$ through a fully-saturated granular media by substituting into Equation 3.24 the average value of R given by ϕ/s. Noting that the discharge Q through an area A of the porous medium normal to i is $\bar{u}_i\phi$, they obtained

$$q_i = \frac{-k}{\mu}\frac{\partial P^*}{\partial x_i}$$

in which

$$k = \frac{\phi^3}{k_s T s^2} \quad , \qquad 3.27$$

and q_i is Q/A. Equation 3.27 is known as the Kozeny-Carmen equation for permeability. Actually, Kozeny combined $k_s T$ in his original equation and determined that it was about 5. Carman pointed out that Kozeny's constant was actually a product of two factors, that is, k_s and T. However, Carman agreed that the product was about 5, since he regarded k_s to be about 2.5 and T to be about 2.0 for sand beds. The equation with these constants has been used in some industries as an indirect method of determining s for granular powders and casting sands.

The same constants have been found for sands by many later investigators, but the relationship fails completely when applied to something other than sands. This is not surprising in view of the discrepancy between $\bar{R}^2$ and $\overline{R^2}$ which was discussed in Section 3.4.1. Also it cannot be applied to media for saturations less than 1.0.

3.6.1 Generalized Kozeny-Carman equation - A version of Equation 3.24 should be valid for porous media in general, provided account is taken of the range of values of R due to the distribution of pore-sizes. A relationship valid for $S < 1$ can be obtained by accounting also for the changes in $\overline{R^2}$ and T with S.

An examination of Equation 3.22 indicates that a more valid expression for $\overline{R_w^2}$ might be obtained by averaging R_w^2 over the range of saturations $(S - S_r)$ that actually contributes significantly to flow. This gives

$$\overline{R_w^2} = \frac{\sigma^2 \cos^2\alpha}{p_d^2 S_e} \int_0^{S_e} (\frac{p_d}{p_c})^2 \, dS_e \quad .$$

The value of the integral is not actually changed significantly by the substitution of S_e for S because the ratio (p_d/p_c) approaches zero for $S < S_r$, but $\overline{R_w^2}$ is made slightly larger because $1/S_e$ is slightly larger than $1/S$. This is because $\overline{R_w^2}$ is averaged over only that part of the pore space in which R has a size large enough to permit significant flow. Observing that q_i for the wetting phase is $\bar{u}_i \phi_e S_e$, and employing the expression for $\overline{R_w^2}(S_e)$ and Equation 3.25 for $T_w(S_e)$, results in

$$q_{wi} = \frac{-k_w(S_e)}{\mu_w} \frac{\partial p^*}{\partial x_i}$$

in which

$$k_w(S_e) = \frac{\sigma^2 \cos^2\alpha \; \phi_e}{p_d^2 k_s T_{1.0}} S_e^2 \int_0^{S_e} (\frac{p_d}{p_c})^2 \, dS_e \quad . \qquad 3.28$$

Similarly,

$$q_{nwi} = - \frac{k_{nw}(S_e)}{\mu_{nw}} \frac{\partial p^*}{\partial x_i}$$

in which

$$k_{nw}(S_e) = \frac{\sigma^2 \cos^2\alpha \; \phi_e}{p_d^2 \; k_s \; T_{1.0}} (1-S_e)^2 \int_{S_e}^{1.0} (\frac{P_d}{P_c})^2 \, dS_e \quad . \qquad 3.29$$

Equations 3.28 and 3.29 are analogous to equations first proposed by Burdine (1952). They may be regarded as generalized Kozeny-Carman equations, a name proposed by Wyllie and Spangler (1952) for somewhat similar equations. They are not valid for non-isotropic media, because the analysis assumes that the pore space has no preferred orientation.

Equations 3.28 and 3.29 were greatly simplified by Brooks and Corey (1964) by substituting Equation (2-9) into the integrals and performing the indicated integration. They obtained

$$k_w = k \quad \text{for} \quad p_c \leqq p_d$$

and

$$k_w = k(\frac{p_d}{p_c})^{\eta} \quad \text{for} \quad p_c \geqq p_d \qquad 3.30$$

in which k is the value of k_w when S_e is 1.0, and

$$\eta = 2 + 3\lambda \quad . \qquad 3.31$$

In terms of S_e, they obtained

$$k_w = k \; S_e^{\varepsilon} \qquad 3.32$$

in which

$$\varepsilon = \frac{2 + 3\lambda}{\lambda} \quad . \qquad 3.33$$

For the non-wetting phase, they obtained

$$k_{nw} = k(1-S_e)^2 \; (1-S_e{}^{\gamma}) \qquad 3.34$$

for $S_e < S_m$, in which S_m is some maximum value of S_e at which k_{nw} exists, and γ is $(2 + \lambda)/\lambda$. The value of S_m is usually about 0.85 for homogeneous and isotropic materials.

Equations 3.30 through 3.34 were verified experimentally by Brooks and Corey (1966) and by Laliberte et al. (1966). Laliberte et al. (1968) obtained an expression for k by integrating Equation 3.28 between the limits of S_e from 0 to 1, to obtain

$$k = \frac{\phi_e \sigma^2 \cos^2 \alpha}{k_s \, T_{1.0} \, p_d^2 \, \gamma} \qquad 3.35$$

in which γ is the same parameter that appears in Equation 3.34. Since they used a light oil (Soltrol*) as a wetting phase in obtaining $p_c(S)$, and since Soltrol has a negligible angle of contact, they set $\cos^2\alpha$ equal to 1. They also accepted the published approximations for k_s and $T_{1.0}$ so that

$$k \cong \frac{\phi_e \sigma^2}{5 \, \gamma p_d^2} \quad . \qquad 3.36$$

Combining Equation 3.36 with Equation 3.32, for example, gives

$$k_w \, (S_e) \cong \frac{\phi_e \sigma^2}{5 \, \gamma p_d^2} \, S_e^{\varepsilon} \quad . \qquad 3.37$$

3.7 PERMEABILITY

The analysis which begins in Section 3.1 by a consideration of the motion of fluid particles leads finally to Equation 3.28 which is a macroscopic flux equation. That is, q_{wi} is a volume flux rate averaged over an element of area that includes solid as well as pore space. The fluid does not actually flow through all of the area, so q_{wi} does not actually represent the seepage velocity $\bar{u}_i$. The relationship between q_{wi} and $\bar{u}_i$ is given by

$$q_{wi} \cong \phi_e S_e \bar{u}_i \quad . \qquad 3.38$$

The relationship between q_{wi} and the velocity $\bar{u}_t$ of particles tangential to their flow path is approximated by

$$q_{wi} \cong \phi_e S_e \frac{\ell}{\ell_e} \bar{u}_t \cong \phi_e \frac{S_e^2}{\sqrt{2}} \bar{u}_t \quad . \qquad 3.39$$

*Manufactured by Phillips Petroleum Co., Bartlesville, Oklahoma.

Consequently, q_{wi} is not a velocity component of the center of mass of a fluid particle. Therefore, Equation 3.28 is more accurately called a flux equation than a "flow" equation or an equation of "motion".

The proportionality function $k_w(S_e)$ in Equation 3.28 is called effective permeability in the petroleum literature. It is a function of the geometric properties of the pore space occupied by the fluid under consideration (wetting fluid in this case). It has the dimensions of L^2. In the petroleum literature, permeability (when used without adjectives) refers to the value of $k_w(S_e)$ when S_e is 1.0. This is often designated simply by k as was done in Equations 3.30 and 3.34.

The ratio $k_w(S_e)/k$ is called relative permeability to the wetting phase in the petroleum literature and is designated by k_{rw}. Likewise $k_{nw}(S_e)/k$ is designated by k_{rnw}.

3.7.1 Factors affecting permeability - According to Equations 3.32 and 3.34, the maximum value of both k_w and k_{nw} is k. There are factors which are not taken into account in the theory, however, which invalidates this conclusion. One of these factors is the reaction of clay minerals to some liquid, especially water with a low electrolytic content. Another is the failure of the continuum theory in respect to gases at ordinary pressures.

Factors which are indicated by Equation 3.37 to affect $k_w(S_e)$ include:

(1) effective porosity,
(2) maximum pore size,
(3) pore-size distribution,
(4) effective saturation.

The shape factor k_s and $T_{1.0}$ probably do not vary over a very wide range, but shape of grains might have an important effect on λ, the pore-size distribution index. The same factors affecting $k_w(S_e)$, of course, also affect $k_{nw}(S_e)$.

Another factor of importance, not accounted for in the capillaric theory, is isotropy. The theory assumes that the media are isotropic so that the pore channels have no preferred orientation and neither k_w nor k_{nw} have directional properties. A related factor is the degree of interconnection of pore space which the theory assumes to be completely random.

3.7.2 Klinkenberg effect - When the fluid is a gas at normal pressures, a fluid particle meeting the requirements for the continuum analysis usually does not exist. This is because, such a particle must be small compared to the pore dimensions and large compared to the mean free path of fluid molecules. With pore sizes of the order of magnitude of those in common earth materials, these requirements are incompatible.

The error resulting from the assumption of a continuum with derivatives of velocity existing at the boundary is usually small in the case of coarse sands, but huge in the case of silts or clays. In the latter case, the difference in k_g as compared to k_w, even when the latter is measured with a liquid that does not react with clays, may be a factor of 2 or 3.

The reason for this huge discrepancy is that the specific surface for fine materials is large, and the boundary condition of a zero velocity at the solid surface is not satisfied in the case of gases unless the mean pressure is, at least, 5 atmospheres. This phenomenon has been called gas slippage and, in respect to flow in porous media, the Klinkenberg effect.

Klinkenberg (1941) measured gas permeabilities as a function of the inverse of mean pressure in a variety of media. He obtained data typified by that shown in Figure 3-8.

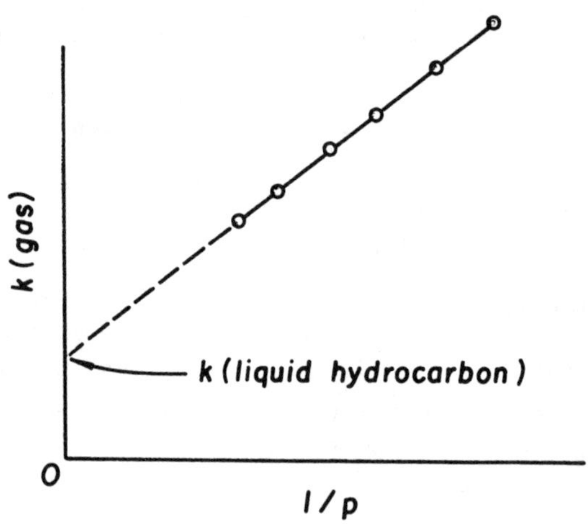

Figure 3-8. Permeability of a gas as a function of the inverse of mean pressure.

Klinkenberg found that the extrapolation of the curve when 1/p is zero gave a value of k_g which was identical to that for hydrocarbon liquids. In the petroleum literature the permeability to gas found by this type of extrapolation is sometimes designated by k, without subscripts, and is called simply permeability. It is regarded as being a property of the pore geometry unaffected by reaction with liquids, e.g., clay swelling.

3.8 DARCY's EQUATION

Despite the obvious shortcomings of the theory upon which Equation 3.28 is based, it is to be expected that an equation of the form

$$q_i = -\frac{k_i}{\mu}\frac{\partial p^*}{\partial x_j}\delta_{ij} \qquad 3.40$$

or its equivalent

$$q_i = -\frac{k_i(\rho g)}{\mu}\frac{\partial H}{\partial x_j}\delta_{ij} \qquad 3.41$$

would be a good approximation for a medium containing only one fluid of constant density. Here δ_{ij} is the Kronecker Delta and the subscript i implies that the medium may not be isotropic so that q_i is a component of $\underline{q}$ in a direction i, and k_i is a permeability associated with this direction. In the most general case k should be regarded as a tensor quantity. However, by aligning the coordinate system with the principle axes in respect to permeability, the k tensor is converted to a diagonal tensor and k becomes a vector quantity in effect. It is assumed here that this is always possible.

Equation 3.40 or 3.41 can be regarded as describing a flux component at a point. If the flux rate q_i is integrated over a large area and $\partial p^*/\partial x_i$ is integrated over a macroscopic portion of the flux path, the result is

$$\frac{Q_i}{A} = -K_i\frac{\Delta H}{L_j}\delta_{ij} \qquad 3.42$$

in which A is an area of a porous system normal to i, K_i is related to k_i by

$$K_i = \frac{k_i \rho g}{\mu} ,$$

averaged over a length of medium L in the direction i, and ΔH is a piezometric head difference measured between two points which are a distance L apart. The coefficient K_i has the dimensions of velocity and is called hydraulic conductivity to distinguish it from k_i which has the dimensions of L^2. In some literature, however, K_i is called permeability and k_i specific or intrinsic permeability. In any case, K_i includes the effect of both fluid and medium properties, whereas k_i theoretically includes only geometric properties of pore space.

If the medium is isotropic, the subscript i can be dropped. The equation is now

$$\frac{Q}{A} = -K \frac{\Delta H}{L} \tag{3.43}$$

in which A is taken to be normal to the direction in which L and ΔH are measured.

Equation 3.43 is called Darcy's law and was discovered by Darcy (1856) experimentally. All other forms of the flux equation for flow in porous media are sometimes called Darcy's law also, but this designation is not entirely valid.

3.8.1 The generalized flux equation - One version of the flux equation which might have even more general validity than Darcy's law is

$$q_i = \frac{k_i}{\mu} \left[- \frac{\partial p}{\partial x_j} + \rho g_j\right] \delta_{ij} \quad , \tag{3.44}$$

in which k_i refers to the permeability of either the wetting or non-wetting phase. Although k_i is treated here as a vector rather than a tensor, Equation 3.44 assumes neither isotropy nor a combined force potential. Furthermore, Equation 3.44 is much more convenient to use when dealing with the flow of gases as well as liquids. The reason is that ρ_w and ρ_g differ by several orders of magnitude. Consequently, it is more convenient to add equations when the potentials are expressed in terms of pressures rather than heads.

In this text, therefore, Equation 3.44 is used whenever the problem deals with the simultaneous flow of mixed fluids. If the media are known to be isotropic and a combined potential exists, it is permissible to write Equation 3.44 in the vector form

$$\underline{q} = \frac{k}{\mu} \left(- \frac{\partial p^*}{\partial x_i} \underline{e}_i\right) \qquad 3.45$$

in which $\underline{e}$ is a unit vector, and the repeated subscripts imply a summation over three orthogonal directions. Note, that the direction of $\underline{q}$ and the negative pressure gradient correspond only when k is a scalar.

3.8.2 Units used in flux equations - Practically every system of units has, at one time or another, been used in connection with flux equations. More often than not, the systems used have been inconsistent. The latter are often erroneously called "practical" units. In order to discourage the use of inconsistent and varied units, only consistent cgs units are used in this text.

A list of units to be used are given in Table 3-1.

The use of an inconsistent system of units in the petroleum industry is so widespread in the literature that it cannot be ignored completely. In the petroleum system, p is expressed in atmospheres/cm^2, viscosity in centipoises, and the other units, except k, are as indicated in Table 3-1. The k that results from this system is inconsistent in that it corresponds to no ordinary unit of length squared. It is called a

Table 3-1. System of units for describing flow in porous media.

Quantity	Symbol	Unit
Force	F	dyne
Mass	M	gram
Length	L	centimeter (cm)
Time	t	second (sec)
Pressure	p	dynes/cm^2
Density	ρ	grams/cm^3
Scalar gravity	g	dynes/gram or cm/sec^2
Vector gravity	$\underline{g}$	dynes/gram or cm/sec^2
Permeability	k	cm^2 or square microns (μ^2)
Conductivity	K	cm/sec
Viscosity	μ	poises (dynes-sec/cm^2)
Volume flux	q	cm/sec
Surface tension	σ	dynes/cm

darcy unit d and has a value of about 0.987×10^{-8} cm^2 or about 0.987 μ^2. Another unit of k in common usage in petroleum and geology literature is the millidarcy md which is 0.001d.

3.8.3 Problem of non-homogeneity - When a differential form of the flux equation is used to describe flux at a point, the question of isotropy arises but not homogeneity. However, when an integrated form of the equation is used, the lack of homogeneity may invalidate conclusions drawn from models that assume homogeneity. This is particularly true in respect to k as a function of saturation when the media are layered [Corey and Rathjens (1956)].

When media are layered, such that the layers are very thin, it is difficult to analyze the individual layers as separate systems. In this case, the assembly of layers will act as an anisotropic system. This may drastically alter $k_{nw}(S_e)$ and to a somewhat lesser extent $k_w(S_e)$.

Consider, for example, a rock consisting of thin layers with contrasting values of p_e, from which samples have been taken using a core drill both parallel and normal to the bedding planes. Sketches of some hypothetical cores of this type are shown in Figure 3-9.

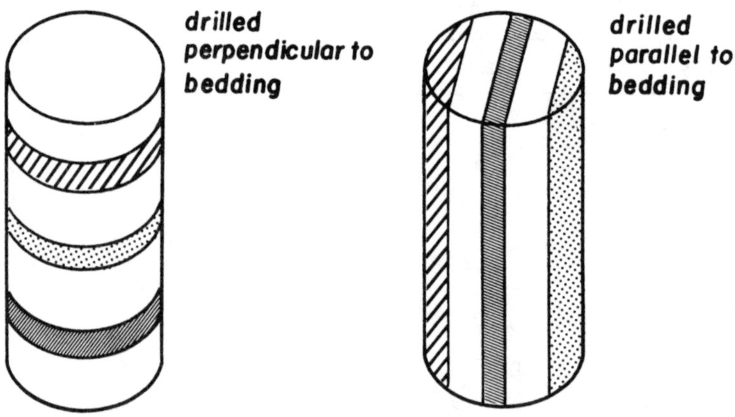

Figure 3-9. Core samples from layered rocks.

In Figure 3-9, the darker layers represent fine materials with a very large p_e. The opposite is depicted as granulated layers. Air permeabilities determined, as a function of average liquid saturation, in an apparatus that permits air flow parallel to the axes of the cores are very different for the two cases and neither resembles $k_{nw}(S_e)$ predicted by Equation 3.34.

In particular, the values of S_m are very small (possibly approaching zero) for the perpendicular core and large (possibly approaching 1.0) for the parallel core. This is in contrast to an average value of S_m which is about 0.85. The values of $k_w(\bar{S}_e)$ also are different from predicted values but the difference may not be quite so extreme as for $k_{nw}(\bar{S}_e)$.

One practical application of the case presented is in respect to measured values of S_m. If $S_m < 0.85$, homogeneities across the flow path should be expected. If $S_m > 0.85$, homogeneities parallel to the flow path are to be expected.

Many other examples of effects from small scale nonhomogeneities can be found, but the case presented should serve to illustrate the possibilities.

3.8.4 The $k_w(S)$ function during drainage - The relationships shown in Figures 2-4 and 2-6 for $p_c(S)$ were obtained by measurements made under essentially static conditions. A question arises as to whether $p_c(S)$ would be the same if the measurements are made while S is continuously changing rather than changing by increments. If there is a difference, it would be expected that $k(S)$ would also change, in view of the theory presented in Section 3.6.1.

Topp et al. (1967) and Corey and Brooks (1975) have presented evidence that such a difference actually exists. The difference arises primarily in the range of $p_c < p_e$. In this range of p_c, both the $p_c(S)$ and $k(S)$ functions may be time-dependent. During a continuous process when an interconnected non-wetting phase does not exist, practically no desaturation takes place, especially if large masses of media are involved which are not close to an exposed boundary. If the media are allowed to remain at one value of $p_c < p_e$ for a long time, however, it may be possible for a gaseous non-wetting phase to evolve from solution or to pass through restricting liquid films by diffusion.

This explanation is speculative, but in any case, both the $p_c(S)$ and $k(S)$ functions are different when measured during a continuous drainage process as Corey and Brooks have shown.

The difference in respect to the $k_w(S)$ function is shown in Figure 3-12, Section 3.8.6. The use of $p_c(S)$ data obtained during a continuous process, however, should not give different values for a calculated λ, because the difference occurs within the range of $p_c < p_e$, and such data are not used in the calculation in any case.

3.8.5 Typical relationships of k(S) - Typical curves of k(S) for both the wetting or non-wetting phases are shown in Figure 3-10.

The k(S) functions shown in Figure 3-10 are expressed as $k_r(S)$. Here k_r means relative permeability which in this case is defined as $k_w/k_{w(max)}$ or $k_{nw}/k_{nw(max)}$, in which the subscript (max) refers to the maximum value of k for the particular phase under consideration. Usually $k_{nw(max)}$ is somewhat larger than $k_{w(max)}$ because of clay swelling and gas slippage. If k_w is determined with a hydrocarbon liquid and values of k_{nw} are determined by extrapolation to infinite mean pressure, $k_{w(max)}$ and $k_{nw(max)}$ would presumably be the same value. Also, if k_w and k_{nw} are both liquids, e.g., brine and oil, the maximum values of k for both phases may be nearly the same.

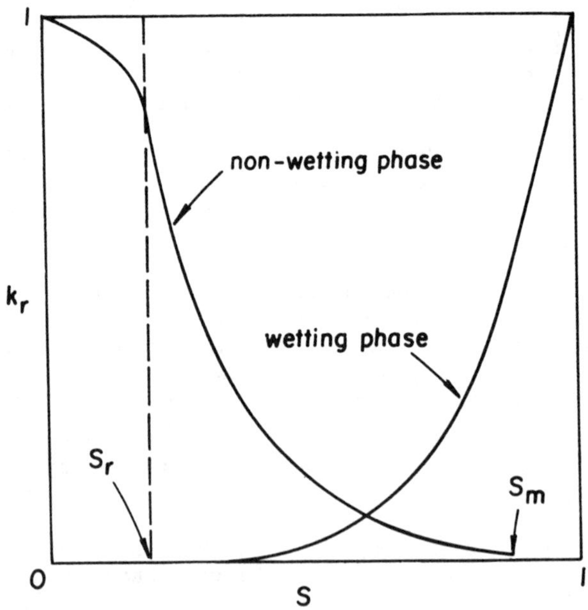

Figure 3-10. Typical relative permeability curves.

Unlike $p_c(S)$ or $k(p_c)$ functions, $k_w(S)$ apparently exhibits very little hysteresis. This is not necessarily the case for $k_{nw}(S)$, however. In particular, the value of S_m may be significantly different when obtained on a drainage cycle as compared to that on a wetting cycle, although this has not been investigated systematically.

Semi-empirical relationships for $k_r(S)$ have been presented by Brooks and Corey (1964) as described in Section 3.6.1. For the wetting phase $k_r(S)$ is expressed by Equation 3.32, that is,

$$k_{rw} = S_e^{\varepsilon} \ .$$

Semi-empirical relationships for $k_w(S)$ have been presented by other authors including:

(1) Childs and Collis-George (1948),

$$K_w = B\theta^3/s^2 \qquad 3.46$$

in which K_w is $k_w \rho_w g/\mu_w$, θ is the volumetric water content ϕS, B is a constant and s is the specific surface.

(2) Irmay (1954),

$$k_{rw} = S_e^{3} \ . \qquad 3.47$$

(3) Corey (1954),

$$k_{rw} = S_e^{4} \ . \qquad 3.48$$

(4) Averjanov (1950),

$$k_{rw} = S_e^{3.5} \ . \qquad 3.49$$

It will be noted, that the relationships of Childs and Collis-George and Irmay could be expressed in the same form, except that the former does not account for the parameter S_r. It is doubtful that any relationship that does not account for such a parameter could be valid for anything except sands with very small values of S_r or some soils that have zero or poorly defined values of S_r.

It should also be noted that Irmay's equation corresponds to that of Brooks and Corey for $\lambda \to \infty$, that is, a uniform pore size. It is significant that experiments which confirm Equation 3.47, e.g., Hausenberg and Zaslavsky (1963), were conducted with sands of uniform grain size.

The relationship of Corey (1954) was determined from experiments with a variety of porous rocks. Note, that it corresponds to Equation 3.32 for a λ of 2, which, according to Brooks and Corey (1964), is a typical value for soil materials as well as porous rocks. Some soils and porous rocks with highly developed structures have values of ε larger than 4.

The experiments of Averjanov leading to Equation 3.49 were made with naturally occurring sand deposits. The value of ε obtained by him has been confirmed by Brooks and Corey for similar materials.

3.8.6 Typical relationships for k(p_c) - Typical curves of $k_{rw}(p_c)$ for a fine sandy soil are shown in Figure 3-11 taken from Brooks and Corey (1964). Note, that as predicted by Equation 3.30, the curves are linear on the log-log plot over a considerable range of p_c. A discrepancy exists on the drainage cycle mainly in the range of $p_c < p_e$. Although, some decrease in S occurs with any increment of $p_c >$ zero, k_{rw} curves show no change of k_{rw} over a significant range of p_c.

Evidently, the initial desaturation that occurs at the boundary does not affect any interconnected flow channels. There is some reduction of k_{rw} just before p_e is reached, although this again is apparently a boundary effect. Evidence for the latter conclusion is presented in Figure 3-12, in which a comparison is shown of data obtained by a continuous drainage process with that of a typical curve determined by steady state procedures.

Note, that in the case of continuous drainage, k_{rw} is not reduced below 1.0 until p_c equals p_e. After this, p_c often actually <u>decreases slightly</u> while k_{rw} is decreasing abruptly. The value of k_{rw} typically decreases by 50 percent, or slightly more, before p_c again reaches p_e. During the drainage period when $p_c \leqq p_e$, the value of p_c is unstable and may undergo small abrupt decreases as air breaks into new regions of the sample. The increase in p_c becomes gradual and stable once p_c again exceeds p_e.

Since White et al. (1970) have shown that air permeability does not exist at $p_c < p_e$, it is reasonable to suppose that

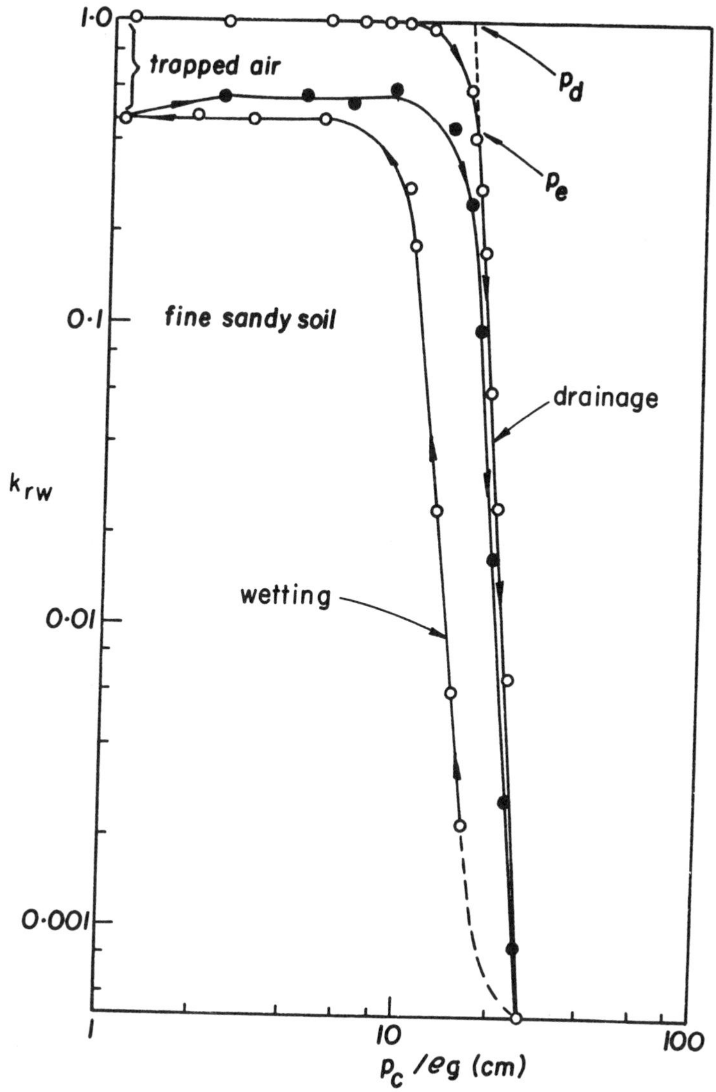

Figure 3-11. Wetting phase relative permeability as a function of capillary pressure on the drainage and wetting cycle.

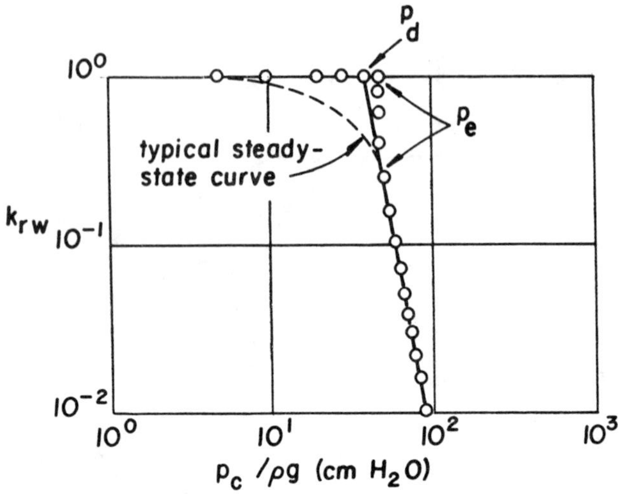

Figure 3-12. Log k_{rw} as a function of log p_c measured during continuous drainage [after Corey and Brooks (1975)].

the erratic behavior of p_c in this range is due to abrupt emptying of certain larger pore channels when air breaks through restricting liquid-filled spaces of somewhat smaller dimensions. Once nearly all of the larger pores contain some air, and this air is interconnected, the unstable behavior stops.

Figure 3-11 also illustrates the effect of hysteresis on $k_{rw}(p_c)$. In the case illustrated, the fine sand was first vacuum saturated and then drained by increasing p_c in increments. At a value of p_c of about 2 p_d, when k_{rw} had decreased to about 5 x 10^{-4}, p_c was decreased in increments until air was entrapped. At that time the rate of increase of k_{rw} was very slow. A gradual increase in k_{rw} did occur with time, however, without any increase in p_c. This phemonenon has been interpreted by Adams et al. (1969) as being due to a slow diffusion of entrapped air out of the system. Adams et al. found that the diffusion of entrapped air out of a system, in which air permeability does not exist, occurs faster from fine materials than from coarser materials.

In reference to Figure 3-11, note that when the value of p_c was again increased with most of the entrapped air still in the soil, k_{rw} did not change until p_e was again reached. The curve then followed practically the original drainage curve rather than following some intermediate curve.

Semi-empirical relationships for $k_w(p_c)$ on the drainage cycle have been proposed by a number of authors including:

(1) Gardner (1958),

$$k_w = a/(b + p_c^n) \qquad 3.50$$

in which a, b and n are constants.

(2) Gardner (1958),

$$k_{rw} = \exp(-ap_c) \quad . \qquad 3.51$$

(3) Arbhabhirama and Kridakorn (1968),

$$k_w = k_{w(max)}/[(p_c/p_d)^n + 1] \quad . \qquad 3.52$$

(4) Brooks and Corey (1964),

$$k_{rw} = 1 \quad \text{for} \quad p_c \leqq p_d$$

and

$$k_{rw} = (p_d/p_c)^\eta \quad \text{for} \quad p_c \geqq p_d \quad . \qquad 3.53$$

The first three empirical relationships for $k_w(p_c)$ are smooth functions that predict a finite decrease in k_w with any finite increase in p_c. In this respect, they do not agree with experimental observations, because k_w remains invariant over a significant range of p_c. With a proper choice of constants, however, the functions can be made to approach, very closely, the actual behavior. Moreoever, the use of a smooth function is convenient for some types of computation.

Equation 3.51 provides the poorest fit to actual data, but it is the easiest to incorporate into mathematical solutions. Consequently, it is the favorite function among applied mathematicians. Neither Equation 3.50 nor Equation 3.51 are dimensionally consistent so that the constants used depend upon the units in which the variables are expressed. Equation 3.52 is much like Equation 3.50 except that the former is dimensionally consistent.

Equation 3.53 cannot be made to fit data on small samples for $p_c < p_e$, but it is realistic to the extent that it predicts a finite range of p_c over which k_w is invariant. Moreover, it comes closer than the other relationships in describing $k_w(p_c)$ for continuous drainage processes.

Brooks and Corey (1966) have also published data for $k_{rnw}(p_c)$. Typical curves of this type are shown in Figure 3.13 for several media representing a wide range of pore-size distributions.

The smooth curves shown in Figure 3-13 were calculated by combining Equation 2.9 and 3.34, that is,

$$k_{rnw} = [1 - (\frac{p_d}{pc})^{\lambda}]^2 \; [1 - (\frac{p_d}{pc})^{2+\lambda}] \qquad 3.54$$

for $p_c \geqq p_e$.

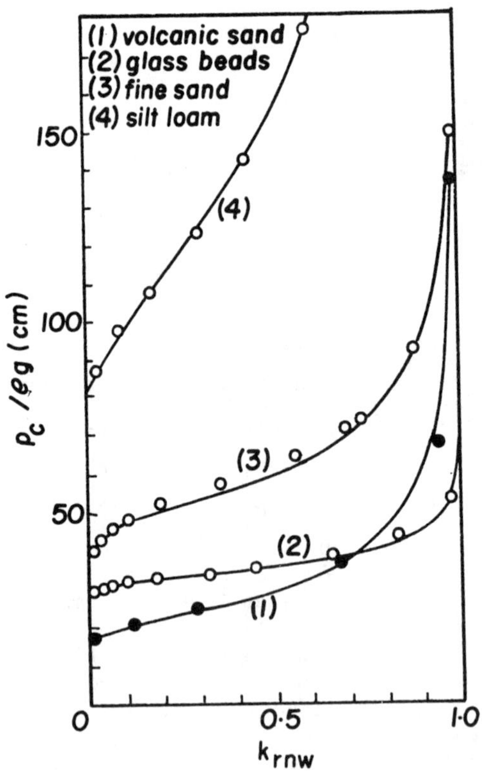

Figure 3-13. Relative permeability of air as a function of p_c compared with theoretical function.

Note that the agreement between theory and measured data is very good considering the highly divergent nature of the materials examined. The materials did have one important property in common, however; that is, they were all quite homogeneous and isotropic.

3.9 SOIL-WATER "DIFFUSION"

For the special case in which p_c is a single-valued function of wetting-phase content, it is possible to write the flux equation,

$$q_{wi} = \frac{k_w}{\mu_w} \left(- \frac{\partial p_w}{\partial x_i} + \rho_w g_i\right)$$

in terms of a gradient of wetting-phase content. Since this procedure is practical only for flux of water in partially saturated soils, the symbols used in this connection are those employed by soil scientists. That is, q_i is a flux of water in the i direction, K is the conductivity of water, $k_w \rho_w g/\mu_w$, and is assumed to be a scalar. The negative gauge pressure head of water is designated as ψ; that is, ψ is $-p_w/(\rho_w g)$.

With these symbols, the flux equation is written as

$$q_i = K\left(\frac{\partial \psi}{\partial x_i} - \frac{\partial h}{\partial x_i}\right) \quad .$$

Assuming ψ to be a single-valued function of θ, that is, ϕS, permits a transformation of variables such that

$$\frac{\partial \psi}{\partial x_i} \equiv \frac{d\psi}{d\theta} \frac{\partial \theta}{\partial x_i} \quad .$$

A function $D(\theta)$, that is also assumed to be single-valued, is defined as

$$D(\theta) \equiv - K(\theta) \frac{d\psi}{d\theta}$$

so that

$$q_i = - D(\theta) \frac{\partial \theta}{\partial x_i} - K(\theta) \frac{\partial h}{\partial x_i} \quad . \tag{3.55}$$

Equation 3.55 was first derived by L. A. Richards (1931). Prior to that time, an analogous equation without the gravity term, was employed on a heuristic basis in recognition of the apparent tendency of water to spread from wetter to drier soil.

It should be recognized, however, that writing the flux equation in the form of Equation 3.55, is a mathematical artifice that results in a loss of generality in the following respects:

(1) The equation is valid only for homogeneous soils such that $\psi(\theta)$ is single-valued.
(2) The air pressure must be atmospheric everywhere, or at least constant, so that $p_w/\rho_w g$ is $-\psi$. Therefore, it cannot be used to describe 2-phase flow phenomena.
(3) An interconnected air phase must exist at all points in the region considered, otherwise, $\psi(\theta)$ is not single-valued.

Practical problems ruled out as applications of Equation 3.55, or at least made inconvenient to solve, include:

(1) flow in layered, or otherwise non-homogeneous soils,
(2) problems of 2-phase flow; for example, it is impossible to analyze what happens as a gas cap breaks into an oil well or what happens when water is driven from a tube of wet soil by air pressure,
(3) infiltration from a ponded source of water, such as, a river bed, canal, reservoir, or even an irrigation furrow,
(4) drainage from a soil immediately above a water table.

Equation 3.55 does have some mathematical advantages, especially for cases of horizontal flow, or when the gravity term is negligible. In such cases, the equation takes the form of a true "diffusion" equation or a "heat" flow equation, for which solutions to many boundary value problems have been tabulated.

The function $D(\theta)$ can also be formulated in terms of the pore-size distribution index of Brooks and Corey as

$$D(\theta_e) = [\frac{K_m \psi_d}{\lambda \phi_e}] \theta_e^{(2+1/\lambda)} \qquad 3.56$$

in which

$$\theta_e = \frac{\theta - \theta_r}{\theta_m - \theta_r} ,$$

$$\psi_d = \frac{p_d}{\rho_w g} ,$$

$$\phi_e = \theta_m - \theta_r \quad ,$$

and

$$K_m = K \text{ at } \theta_m, \text{ the maximum field water content.}$$

3.10 MEASUREMENT OF RELATIVE PERMEABILITY

Methods of measuring permeability of a fully saturated porous sample are discussed in texts on soil mechanics and groundwater hydraulics and are not reviewed here. Measurement of k_r as a function of either p_c or S is a more complex problem but is fundamental to the development of the mechanics of mixed fluids in porous media.

An extensive review of those methods which have been used by soil scientists has been given by Klute (1972). A review of a few of the more common procedures used to determine $k_r(S)$ or $k_r(p_c)$ on rock cores has been presented by Richardson (1961). The methods can be classified as either steady-state or unsteady-state procedures, but there are innumerable variations of each. Only a few of the simpler procedures are reviewed here.

Some procedures can be understood only with additional background theory relative to steady and unsteady flow. A further discussion of such methods is presented in Chapters IV and V as applications of the theory presented in those chapters.

3.10.1 Steady-state methods - The variables: pressure, saturation and flux rate are held constant with respect to time, at particular values of pressure and saturation, until the measurement of k_{rw} and k_{rnw} are completed. In some cases only k_{rw} (or k_{rnw}) is determined, but in other cases both k_{rw} and k_{rnw} are measured for a particular p_c and S.

With some procedures, p_c and S are permitted to vary along a column and k_r is determined at various points along the column. More often, a uniform p_c or S is established throughout a sample or, at least, within a test section of a column during the determination of particular values of k_{rw} and k_{rnw}. Methods of achieving a uniform p_c and S within a sample include:

(1) Downward flow of a liquid wetting phase is allowed under a gravitational force only. The non-wetting

phase is air at atmospheric pressure, but k_{rg} is not determined. The first measurements of $k_{rw}(p_c)$ (of which there is a record) were made using this system by L. A. Richards (1931).

(2) Simultaneous flow of two fluid phases is produced under the same pressure gradient. This procedure was introduced in the petroleum industry by Hassler et al. (1944), for determination of k_{rw} and k_{rnw} on small cores of porous rock. The method was adapted for use on soil samples by Brooks and Corey (1966).

(3) Two fluid phases are injected simultaneously through a porous sample, with the individual flow rates controlled independently, rather than the pressure gradients being controlled. This is called the Penn State method and has been used frequently in the petroleum industry. It gets its name as a result of its having been developed in the Petroleum Engineering Department at Pennsylvania State University.

(4) Upward flow of air through a sample is induced by a pressure gradient equal to the pressure gradient in a static liquid phase. This method was designated as the stationary liquid method by Osoba et al. (1951). It is, of course, useful for the determination of k_{rg} only, but it is very convenient and precise for that purpose.

3.10.2 The Richards method - An apparatus similar in principle to that invented by Richards (1931) is shown in Figure 3-14.

By adjusting the pressure at the upper and lower barriers to compensate for the head loss through the barriers (and contact regions), the pressure difference in the sample, as recorded over the length L, can be adjusted to zero. Under this condition the wetting liquid flows under the influence of gravity only, and the non-wetting phase (in this case air) is everywhere at atmospheric pressure. In order to be sure that the air is actually at atmospheric pressure, air vents must be provided which, at the same time, must not permit significant evaporation of liquid from the sample.

The outflow rate is measured at the same time the pressures at the upper and lower tensiometer rings are recorded, so that Darcy's equation can be solved over the length L. Note that the tensiometers are at some distance inside the ends of the sample to avoid "end effects". The tensiometers are designed to fit around the circumference of the sample so that they do not distort the normal flow path through the sample.

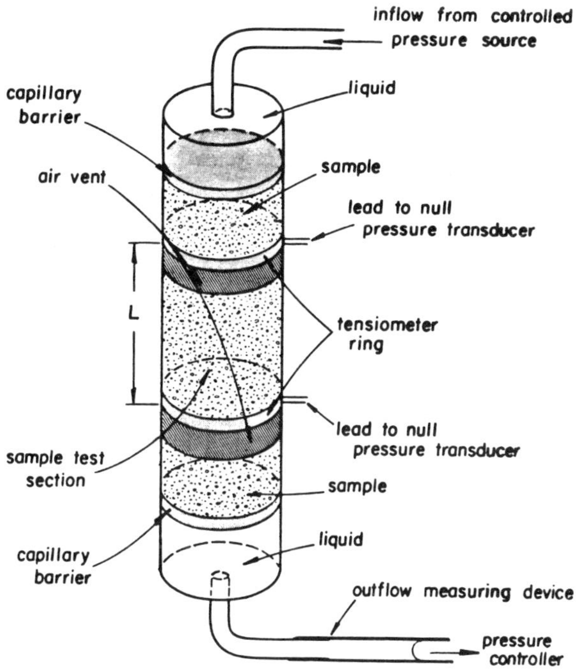

Figure 3-14. Richards permeameter.

Pressures at the top and bottom barriers are changed in increments, maintaining a zero pressure gradient at each increment until a steady state is achieved at which time a value of k_w and k_{rw} are measured. The corresponding values of p_c are determined by the difference between atmospheric and liquid pressures, the latter being less than atmospheric after desaturation occurs.

If it is desired to determine S as well as p_c, this can be done by scanning the test section of the sample with gamma attenuation equipment.

When selecting capillary barriers to be used for particular samples, it is important to select materials that have as much permeability as possible, consistent with the requirement that the barrier p_e not be exceeded in the process of desaturating the sample. It is not practical to use the same kind of barrier material for all types of samples. It is also important to use clean liquids, free of algae and other micro-organisms to minimize plugging of the capillary barriers. When the barriers have a high resistance compared to that of the sample, it is extremely difficult to exercise control over the pressure gradient in the sample.

With an appropriate change in design details, the Richards apparatus can be adapted for use with both rock cores and undisturbed soil samples [Laliberte and Corey (1967)]. It is much easier, however, to obtain precisely reproducible results using hydrocarbon liquids, e.g. Soltrol oil, than with water as the wetting fluid. This is because hydrocarbons have more consistent wetting properties and surface tensions that are less likely to become contaminated with algae or to cause progressive changes in media structure due to clay dispersion.

3.10.3 Long and short columns - It is possible to obtain a more or less uniform p_c and saturation in the upper part of a long column during steady downward flow to a "water table" located near the base of the column. This can be done without any capillary barriers at the top or bottom and without any adjustment other than that necessary to control the inflow rate at the top of the column.

The long-column technique was first suggested by Childs and Collis-George (1950). It is useful for unconsolidated materials that can be packed uniformly in a long column. Non-homogeneities will prevent the establishment of a uniform p_c and S in any part of the column.

A uniform p_c and S can be established during downward flow in a much shorter column by creating a high p_c at its base rather than maintaining a zero p_c at that point.

The distribution of p_c and S during steady downward flow in porous media columns is considered again in Chapter IV as an application of steady flow theory. Some of the pitfalls in the operation of both long and short column permeameters are discussed in the context of theory presented in that chapter.

3.10.4 Air relative permeability measurements - By far, the simplest method of determining air permeability as a function of liquid saturation is the <u>stationary-liquid</u> procedure. With this method air flow is upward through a confined sample of porous medium containing a static liquid. The pressure gradient producing air flow is made exactly equal to the pressure gradient caused by the weight of the static liquid so that a uniform p_c and S are established during the measurement of k_g. The corresponding value of S is obtained by weighing the sample before and after the k_g measurement. An apparatus used for this purpose is illustrated in Figures 3-15, 3-16 and 3-17.

With this apparatus, a cylindrical sample is confined laterally by a rubber sleeve which fits inside an outer cylinder. When the sample is in place, the sleeve is pressed against the sample by air pressure. If the sample is a rigid rock core, the pressure should be about 7 atmospheres. If an unconsolidated sample is used, a pressure of 1/3 atmosphere is sometimes sufficient to prevent bypassing of air.

The top and bottom of the sample must be open to air flow, of course, and in the case of unconsolidated samples, this requires some type of low resistance screen to support the sample at the bottom. Provision must be made to change the liquid saturation between measurements of k_g. With rock samples, this is easily done by blotting the sample with an absorbent tissue or simply allowing the liquid to evaporate until a desired weight is obtained. In the case of unconsolidated samples, the problem of desaturation is more complex, and a number of procedures have been devised. Brooks and Corey (1966), for example, employed a sample that was packed into an annular space between an outer pressurized sleeve and a hollow porous cylinder through which the liquid could be extracted. The saturation was computed by a material balance procedure.

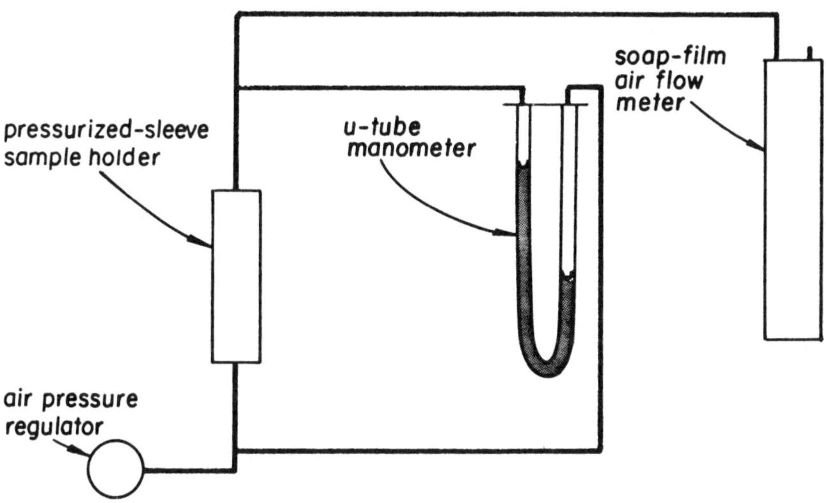

Figure 3-15. Air permeameter with pressurized-sleeve sample holder and soap-film air flow meter.

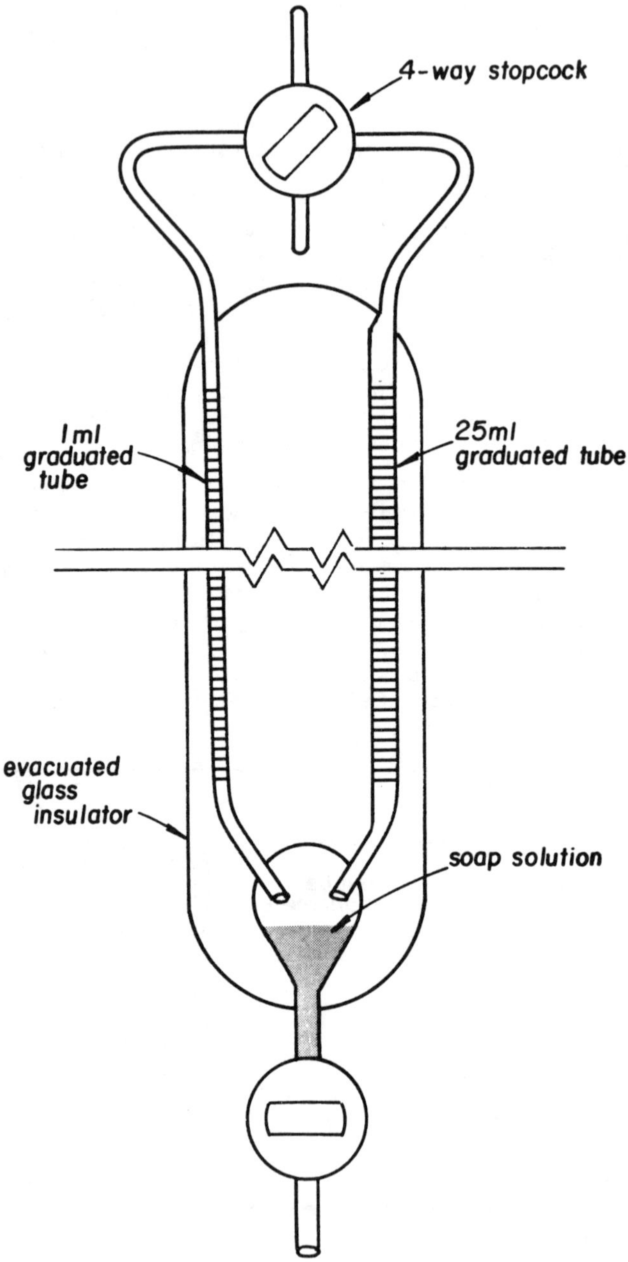

Figure 3-16. Soap-film flow meter.

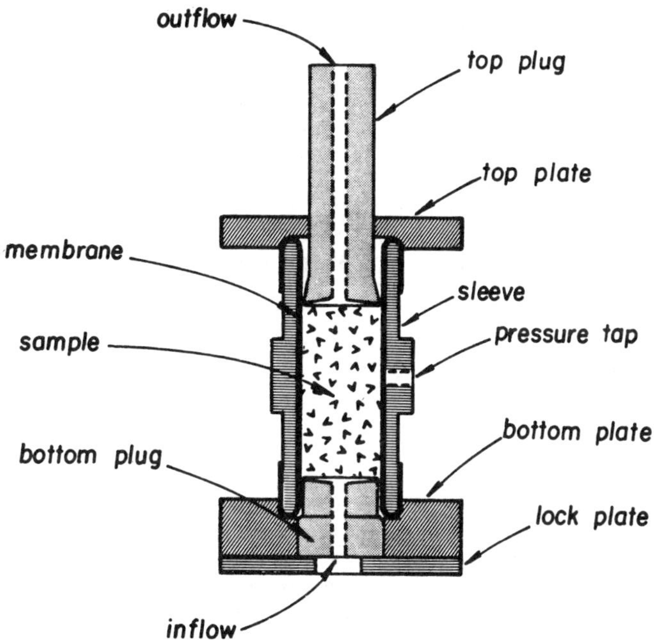

Figure 3-17. Pressurized sleeve sample holder.

The measurement of flux rate for k_{rg} measurements requires a method for measuring very small flux rates precisely. This is best accomplished with a soap-film flow meter using at least two calibrated tubes of varying diameters. The smaller tube should permit the precise measurement of flux rates at least as small as 10^{-3} cm^3/sec. A soap-film tube made from a 1 ml pipette is satisfactory for the small flow rates whereas a 25 ml or larger pipette is useful for the larger flow rates. These suggested sizes are for sample cross sections of 5 to 20 cm^2.

The flux rates cannot be varied by changing the pressure difference causing flow because the air pressure gradient must be held equal to the static pressure gradient in the liquid, in order to maintain a uniform p_c and S in the sample.

3.10.5 Unsteady-state methods - Usually, no attempt is made to induce a uniform p_c or S within a test section of a sample during a measurement. Also, the pressure, saturation, and flux rates are permitted to vary with time at all points within a sample. The computation of k_{rw} and k_{rnw} from measurements made during such experiments are usually indirect and depend upon the solution of unsteady flow equations.

The best known procedure under this classification is the Welge (1952) technique which depends upon an integration of the Buckley-Leverett displacement equation. With this method, a wetting fluid is displaced from a sample by a non-wetting fluid under an extremely large pressure gradient. The pressure gradient is large enough to permit the assumption that the two fluids flow through the sample under the same pressure gradient, even though the saturation is nonuniform. Values of k_{rw} and k_{rnw} are determined from the instantaneous rates at which the two fluids emerge from the outflow end of the sample. The corresponding values of S are those calculated for the outflow end of the sample by the Welge integration. This subject is discussed in Chapter V as an application of unsteady flow theory.

Other procedures which come under this classification include variations of the outflow method, first described by Gardner (1956). With these procedures, a saturated soil sample is placed in a capillary pressure cell and desaturated by applying increments of p_c. In the original version, outflow was in response to very small increments. Doering (1965) modified the procedure by desaturating samples in one-step increments. In any case, values of soil water diffusivity D are obtained indirectly by applying unsteady flow theory. Values of k_{rnw} are not obtained.

Outflow methods have not given consistently reproducible results because the resistance of the region of contact between sample and capillary barrier varies erratically with the p_c applied. Attempts have been made to account for this resistance upon the assumption that the resistance is proportional to the flux rate, but this assumption is not even close to being satisfied. Air diffusing through the capillary barrier and reappearing as bubbles in the outflow system also causes problems.

A procedure called the continuous drainage method by Corey and Brooks (1975) is an exception among the unsteady-state procedures in that more or less uniform (although continuously decreasing) saturation is produced in a portion of a draining column where observations of p_c and S are made. Like other column procedures, downward flow is produced under the influence of gravity only. This procedure might also be called a "pseudo-steady" method because the pressure distribution in the column is nearly like that in a column during steady downward flow. The theory of this behavior is also discussed in Chapter IV.

3.11 NON-DARCY FLOW

From the theoretical considerations leading to the generalized Kozeny-Carman equation, it is expected that a linear flux equation would be invalid for velocities such that fluid inertia becomes a factor. Although the convective acceleration term in the equation of fluid motion,

$$u_j \frac{\partial u_i}{\partial x_j} ,$$

should be statistically zero when integrated over a macroscopic area and distance, the local effect of such acceleration may be to create a spatial distribution of velocity different from that predicted by ignoring inertia.

In this case, the shear force,

$$\mu \frac{\partial^2 u_i}{\partial x_j \, \partial x_j} ,$$

may be increased non-linearly with respect to $\bar{u}_i$, with a corresponding non-linear increase in $\partial p^* / \partial x_i$. Note that this is not equivalent to saying that turbulence develops in the porous medium.

By analogy with flow through large conduits, a Reynolds number (Re) is often used as a criterion to define the regime for which inertial forces are unimportant. According to this analogy

$$\text{Re} \equiv \frac{gD\rho}{\mu} ,$$

in which D is some length parameter characterizing the pore geometry. It would appear that D should be related to the largest pore size, perhaps σ/p_d would be a reasonable choice, but innumerable parameters having the dimension of length have been suggested by various authors. Among these are:

(1) Collins (1961), $D = (k/\phi)^{1/2}$,

(2) Ward (1964), $D = k^{1/2}$,

but by far the most often used is some D determined by a sieve analysis. For example, the 10% size or the 50% size, that is, the grain size that exceeds the diameter of these percentages by weight, have often been employed.

In practically all cases when Re is based on a grain diameter, Darcy's law is valid unless Re exceeds an Re between 1 and 10. This removes the problem of inertial effects from the type of applications considered in this text. However, if non-Darcian flow of this type is suspected, it can be verified by plotting q_i as a function of $\partial p^*/\partial x_i$. If the flow is Darcian, the data should plot as a straight line. If q deviates towards lower values than a linear relationship indicates, inertial effects may be suspected.

A variety of equations have been proposed to describe non-linear flow of the latter type. One of the first, proposed by Forchheimer (1901), is of the form

$$\frac{\partial p^*}{\partial x_i} = aq_i + bq_i^2$$

in which a and b are constants. A verification of an equation of this form has been presented by Ahmad and Sunada (1969). When equations of this form are used, the constant b is small so that when q_i is small, the equation reduces to Darcy's law.

Another class of non-Darcian flow may occur at very small velocities. It has been theorized that such flows could result from one of several causes, such as:

(1) adsorptive force fields near solid surfaces that interfere with the normal Newtonian behavior of water,
(2) electrostatic force fields set up in the boundary layer as a result of flow tangential to the boundary (i.e., streaming potentials) which act to oppose the flow,
(3) diffusion or other molecular transport processes that act concurrently with convection and which may respond to a potential gradient different from that relating to convection.

Many cases of so-called non-Darcian flow reported in the literature are undoubtedly due to faulty laboratory techniques or misinterpretation of the measurements made. One example, is the use of a piezometric potential for cases in which the density is not constant, or the use of some other so-called "total" potential in an attempt to describe transport resulting from a variety of unrelated mechanisms [Corey and Kemper (1961)].

A review of non-Darcian flow, including a discussion of experimental "pitfalls" leading to misinterpretation of observed behavior, has been presented by Kutilek (1972). Other authors who have written extensively on the subject of non-Darcian flow include Swartzendruber (1962) and Bolt and Groenvelt (1969).

3.12 POTENTIAL FLOW

In groundwater systems (and some petroleum reservoir situations), it may be possible to regard the flow system, on a macroscopic basis, as being irrotational. This requires the aquifer to be homogeneous, isotropic and to have a constant saturation with a homogeneous fluid. Under such conditions it may be possible to define a velocity potential as the scalar -KH and a conjugate function called a stream potential. It may also be possible to assume that the divergence of the flow is zero.

If these condition are satisfied, many flow problems can be solved by the application of potential flow theory. This subject is discussed at length in most books on flow in porous media and groundwater hydraulics. However, for the case of heterogeneous fluids, the saturation is not necessarily constant so that the divergence is, in general, not zero. Furthermore, the saturation depends on the fluid pressure so that potential flow theory has a limited usefulness for the applications analyzed in this text. Consequently, the subject of potential flow is not considered here.

REFERENCES

Adams, K. M., Bloomsburg, G. L. and Corey, A. T. (1969). Diffusion of trapped gas from porous media. Water Resources Research, Vol. 5, No. 4, August, pp. 840-849.

Ahmad, N. and Sunada, D. K. (1969). Nonlinear flow in porous media. J. of Hyd. Div., ASCE, Vol. 05, No. HY6, Proc. Paper 6883, November, pp. 1847-1857.

Arbhabhirama, A. and Kridakorn, C. (1968). Steady downward flow to a water table. Water Resources Research, Vol. 4, pp. 1249-1257.

Averjanov, S. F. (1950). About permeability of subsurface soils in case of incomplete saturation. Engineering Collection, Vol. VII, as quoted by P. Ya, Polubarinova Kochina, The Theory of Ground Water Movement, English translation by J. M. Roger De Wiest, 1962, Princeton University Press.

Bolt, G. H. and Groenevelt, P. H. (1969). Coupling phenomena as a possible cause for non-Darcian behavior of water in soil. Bull. I.A.S.H., No. 2, Vol. 14, No. 2, pp. 17-26.

Brooks, R. H. and Corey, A. T. (1964). Hydraulic properties of porous media. Colorado State University Hydrology Paper No. 3, March, 27 p.

Brooks, R. H. and Corey, A. T. (1966). Properties of porous media affecting fluid flow. ASCE, J. Irrig. Drain. Div., IR2, Vol. 92, pp. 61-88.

Burdine, N. T. (1952). Relative permeability calculations from pore-size distribution data. Trans. AIME, Vol. 198, pp. 71-77..

Carman, P. C. (1937). Fluid flow through a granular bed. Trans. Inst. Chem. Eng., London, Vol. 15, pp. 150-166.

Childs, E. C. and Collis-George, N. (1948). Soil geometry and soil-water equilibria. Disc., Faraday Society, No. 3, pp. 78-85.

Childs, E. C. and Collis-George, N. (1950). The permeability of porous materials. Proc. Royal Society, London, Vol. 201, pp. 392-405.

Collins, R. E. (1961). Flow of Fluid Through Porous Materials. Reinhold Publishing Corp., New York, 270 p.

Corey, A. T. (1954). The interrelation between gas and oil relative permeabilities. Producer's Monthly, Vol. XIX, No. 1, November, pp. 38-44.

Corey, A. T. and Brooks, R. H. (1975). Drainage characteristics of soils. SSSA Proc., Vol. 39, No. 2, March-April, pp. 251-255.

Corey, A. T. and Kemper, W. D. (1961). Concept of total potential in water and its limitations. Soil Science, Vol. 91, No. 5, May, pp. 299-302.

Corey, A. T. and Klute, A. (1985). Application of the potential concept to soil water equilibrium and transport. SSSA Journal, Vol. 49, No. 1, pp. 3-11.

Corey, A. T. and Rathjens, C. H. (1956). Effect of stratification on relative permeability. Journal of Petroleum Technology, Trans. AIME, Technical Note 393, December, pp. 69-71.

Darcy, H. (1856). Les fontaines publiques de la ville de Dijon. Victor Dalmint, Paris.

Doering, E. J. (1965). Soil water diffusivity by the one-step method. Soil Science, Vol. 99, pp. 322-326.

Edlefsen, N. E. and Anderson, A. B. C. (1943). Thermodynamics of soil moisture. Hilgardia, Vol. 15, 298 p.

Fair, G. M. and Hatch, L. P. (1933). Fundamental factors governing the streamline flow of water through sand. J. Amer. Water Works Assoc., Vol. 25, pp. 1551-1565.

Forchheimer, P. (1901). Wasserbewegung durch boden. Zeitschrift Des Verines Deutsch Ing., No. 49, pp. 1736-1749, continued in No. 50, pp. 1781-1788.

Gardner, W. R. (1956). Calculation of capillary conductivity from pressure plate outflow data, SSSA Proc., Vol. 20, No. 3, pp. 317-320.

Gardner, W. R. (1958). Some steady state solutions of the unsaturated moisture flow equation with application to evaporation from a water table. Soil Science, Vol. 85, No. 4, pp. 228-232.

Hassler, G. L., Brunner, E. and Deahl, T. J. (1944). Role of capillarity in oil production. Trans. AIME, Vol. 155, pp. 153-172.

Hauzenberg, I. and Zaslavsky, D. (1963). The effect of size of water stable aggregates on field capacity. Department of Civil Engineering, Technion, Haifa, P.H., Vol. 35 as quoted by Bear (1972), pp. 483-487.

Hubbert, M. K. (1940). The theory of groundwater motion. J. Geol., Vol. 48, No. 8, Part I, pp. 785-943.

Irmay, S. (1954). On the hydraulic conductivity of unsaturated soils. Trans. Amer. Geophys. Union, Vol. 35, pp. 463-467.

Klinkenberg, L. J. (1941). The permeability of porous media to liquid and gases. Amer. Petrol. Inst. Drilling Prod. Pract, pp. 200-212.

Klute, A. (1972). The determination of the hydraulic conductivity and diffusivity of unsaturated soils. Soil Science, Vol. 113, No. 4, pp. 264-276.

Kozeny, J. (1927). Über kapillare leitun des wassers im 1 boden, sitzungsber akad. wiss, wien 136, 271-306. (Citation is from a translation by W. F. Striedieck and C. M. Davis, Published by the Petroleum Branch of AIME.)

Kutilek, M. (1972). Non-Darcian flow of water in soils-laminar region. Fundamentals of Transport Phenomena in Porous Media, IAHR, American Elsevier, pp. 327-340.

Laliberte, G. E., Brooks, R. H. and Corey, A. T. (1968). Permeability calculated from desaturation data. Jour. of Irr. and Drain. Div., ASCE, Vol. 94, No. IR1, Proc. Paper 5843, March, pp. 57-71.

Laliberte, G. E. and Corey, A. T. (1967). Hydraulic properties of disturbed and undisturbed soils. ASTM, Permeability and Capillarity of Soils, Special Technical Pub. No. 417, pp. 56-71.

Laliberte, G. E., Corey, A. T. and Brooks, R. H. (1966). Properties of unsaturated porous media. Colorado State University Hydrology Paper No. 17, November, 40 p.

Osoba, J. S., Richardson, J. G., Kerver, J. K., Hafford, J. A., and Blair, P. M. (1951). Laboratory measurements of relative permeability. Trans. AIME, Vol. 192, pp. 47-56.

Purcell, W. R. (1949). Capillary pressures-their measurement using mercury and the calculation of permeability therefrom. Journal of Petroleum Technology, Vol. 1, No. 2, pp. 39-46.

Richards, L. A. (1928). The usefulness of capillary potential to soil-moisture and plant investigators. Journal of Agricultural Research, Vol. 37, No. 1, pp. 719-742.

Richards, L. A. (1931). Capillary conduction of liquids through porous mediums. Physics 1, pp. 318-333.

Richardson, J. G. (1961). Flow through porous media. Handbook of Fluid Dynamics, Section 16, edited by V. I. Steeter, McGraw-Hill Book Co., Inc., New York, pp. 16-3 through 16-112.

Sullivan, R. R. and Hertel, K. L. (1942). The permeability methods for determining specific surface of fibers and powders. Advances in Colloid Science, Vol. 1, Interscience, New York, pp. 37-80.

Swartzendruber, D. (1962). Non-Darcy flow behavior in liquid-saturated porous media. Journal of Geophysical Research, Vol. 67, No. 13, pp. 5205-5213.

Topp, G. D., Klute, A. and Peters, D. B. (1967). Comparison of water content-pressure head data obtained by equilibrium, steady state, and unsteady-state methods. SSSA Proc., Vol. 31, pp. 312-314.

Ward, J. C. (1964). Turbulent flow in porous media. Proc. ASCE, Vol. 90, No. HY5, pp. 1-31.

Welge, H. J. (1952). A simplified method for computing oil recovery by gas or water drive. Trans. AIME, Vol. 195, pp. 91-98.

White, N. F., Duke, H. R., Sunada, D. K., and Corey, A. T. (1970). Physics of desaturation in porous materials. J. Irrig. Drainage Div., Proc. ASCE, IR2, June, pp. 165-191.

Wyllie, M. R. J. and Gardner, G. H. F. (1958). The generalized Kozeny-Carman equation 11. A novel approach to problems of fluid flow, World Oil Prod. Sect. 210-228.

Wyllie, M. R. J. and Spangler, M. B. (1952). Application of electrical resistivity measurements to problems of fluid flow in porous media. Bull. Amer. Assoc. Petrol. Geol., Vol. 36, No. 2, pp. 359-403.

PROBLEMS AND STUDY QUESTIONS

1. If u_i is to be regarded as an analytic function of the space coordinates even at a solid boundary, what does this imply about the magnitude of u_i at a solid boundary? Explain.

2. If u_i does not vary with time at particular points in a system, what does this imply about the acceleration of fluid particles? Explain.

3. Shear acts throughout a fluid volume. Explain why, in evaluating the force per unit volume acting on a fluid particle due to shear, it is necessary to sum the shearing forces only on the boundaries of the particle.

4. Although the direction of shearing forces on opposite faces of a fluid particle are often in opposite directions, the direction of shear at solid boundaries is always in a direction opposite to u_i of the particle. Explain.

5. Despite the fact that ρ in air varies substantially at different elevations within the Earth's atmosphere, it is often possible for atmospheric physicists to define a combined potential, including pressure and gravitational components, for analyzing motion of air masses. This is called the barotropic case. Explain.

6. It is well known that when two water solutions with different concentrations of salts are separated by a membrane that excludes the salt, water will pass through the membrane toward the solution with the higher salt concentration until the pressure is sufficiently higher on that side to counteract the higher osmotic concentration. What can one conclude from this concerning pore size in the membrane relative to the size of fluid particles? What will happen if the pore sizes are made substantially larger, say the size of a pin hole? Explain.

7. If one attempts to derive Poiseuille's equation by regarding the entire volume of fluid held in a section of tube as a free body, and balancing shear and pressure forces on this body, what additional assumption is necessary as compared to when forces are balanced on fluid particles?

8. State as simply as possible the mathematical reason why the Kozeny-Carman equation is not valid for porous media having a range of pore sizes.

9. According to the Kozeny-Carman equation, if one were to mix a teaspoonful of montmorillonite with 10 liters of sand, the value of k calculated might be reduced by an order of magnitude. Explain why this is probably unrealistic.

10. Derive an equation for k_{rnw} as a function of p_c and λ from the "generalized" Kozeny-Carman equation.

11. Would you expect the generalized Kozeny-Carman equation to give better results in terms of "relative permeabilities" or in terms of permeability per se? Explain.

12. Consider a case of steady downward flow of water through a column of homogeneous soil. The system has the following properties:

    (1) The pressure in the water is uniform.
    (2) The porosity of the soil is 0.5.
    (3) The saturation of the soil is 0.7.
    (4) The residual saturation is 0.3.
    (5) When fully saturated, k_w for the soil is 10^{-8} cm^2.

    If dye is injected into the column at some point, estimate the time for some of this dye to appear at a point one meter lower in the column. What if any effect would molecular, or other dispersion phenomena, have on the estimated time of arrival.

13. Estimate a <u>mean</u> distance that dye would have traveled upon arrival at a point 1 meter from the starting point. Assume that the dye molecules are too large to be affected significantly by molecular dispersion.

14. The U.S. Salinity Laboratory at Riverside, California once proposed an index of soil structure stability obtained from the ratio of k_w (when fully saturated with water) to k_g of a dry sample. Explain why such a ratio might not be a measure of stability of soil structure only.

15. Using the generalized Kozeny-Carman equation, derive an expression for k_{rw} a a function of S for a sand having a completely uniform pore-size and a residual saturation of zero. What would be the theoretical values of ε and λ for this case?

16. A sample of porous medium has a value of k_o (to oil) of 0.5 μ^2 as measured in a laboratory at sea level. Estimate the value of k_o (to the same oil) if the measurement is made in a laboratory at a site where the temperature is colder by 10°C and g is 1 percent less. Estimate the conductivity K to an oil with $\mu = 2.5$ centipoises and $\rho = 0.75$ gm/cm^3. Would you expect the permeability to water to be equal to, more than, or less than 0.5 μ^2? What about the permeability to air?

17. Adams et al. (1969) found that entrapped gas diffuses out of samples of fine-textured porous materials (in contact with a source of wetting liquid) faster than from coarser-textured porous materials. Give a theory to explain this behavior.

18. Oil is often forced out of porous rocks by the injection of brine. Explain why a flux equation in the form of the "diffusion" equation, that is, Equation 3.55, is not suitable for analyzing the penetration of brine into the rock in this case.

19. Consider a case in which a porous rock sample containing water at a saturation of 0.72 is placed in a permeameter (of the type shown in Figure 3-15) to measure k_g. If the sample is 5 cm in length and U-tube manometer contains oil with a specific gravity of 0.75, what should be the Δh reading on the manometer to be sure that the brine saturation remains uniform along the length of the sample during the measurement?

20. Describe an experiment to check whether or not a flux equation in the form of Darcy's law or Forchheimer's equation should be used to describe a particular flow system.

CHAPTER IV

STEADY FLOW IN HETEROGENEOUS FLUID SYSTEMS

4.1 FLOW IN A HETEROGENEOUS FLUID SYSTEM

Steady flow implies that the variables p, u, ρ and S do not vary with time at any point within a system under consideration. When this situation exists, the relationship among the variables can be analyzed by employing equations that do not include time as a variable. However, it may be more informative to consider first a more general case in which the variables are not necessarily constant with respect to time and to examine steady flow as a special case of the more general situation.

To analyze the more general case, it is necessary to combine a flux equation with a continuity equation which involves time as a variable.

4.1.1 Continuity for flow in a heterogeneous fluid system

- A continuity equation is one which expresses the conservation of mass for a reference volume element. In some cases an entire system is considered as a reference element. In other cases only a fluid particle is considered. For flow in heterogeneous fluid systems, however, it is usually more informative to consider a volume element of the porous medium such as is used to define porosity or saturation at a point.

When a fluid particle is used as a reference element in fluid mechanics, a continuity equation is written as

$$\frac{\partial(\rho v_i)}{\partial x_i} = - \frac{\partial \rho}{\partial t}$$

which states that the divergence of the mass flow is equal to the rate of decrease of density of the fluid. This equation could also be written in respect to a fluid element within a porous medium, but it does not supply the information needed to solve the type of problems considered here.

Such an equation is useful for situations in which the entire flow region remains completely occupied with one fluid. This is not true for a porous medium in the general case. In fact, when changes in saturation occur, these are usually of much greater significance than changes in fluid density or medium porosity. Consequently, for heterogeneous fluid systems it is usually a valid approximation to ignore changes in density or medium porosity with respect to time.

Employing this approximation, a continuity equation is written for flow of a particular fluid phase in a two fluid system as

$$\frac{\partial q_{wi}}{\partial x_i} = - \phi \frac{\partial S}{\partial t}$$

or

$$\frac{\partial q_{wi}}{\partial x_i} = - \frac{\partial \theta}{\partial t} , \qquad 4.1$$

for which the reference element is a volume element of porous medium. Equation 4.1 is written in respect to the flux of the wetting phase, but it could be written in respect to q_{nw} by noting that

$$\frac{\partial q_{wi}}{\partial x_i} = - \frac{\partial q_{nwi}}{\partial x_i} ,$$

for a case in which compressibility is not a factor.

4.1.2 Simultaneous flow of two fluid phases - Flow equations are a result of combining flux equations with continuity equations. For two fluid phases, the equations are written for each phase separately and then combined. The assumptions are:

(1) The differential form of Darcy's law applies independently to each fluid phase.
(2) Medium and fluid properties, e.g., ϕ, k, ρ and μ are practically constant in time and space.
(3) Permeability can be treated as a scalar.
(4) Compressibility is not an important factor so that

$$\frac{\partial q_{wi}}{\partial x_i} = - \frac{\partial q_{nwi}}{\partial x_i} .$$

(5) The fluids are immiscible so that

$$p_c = p_{nw} - p_w .$$

Based on these assumptions, it is possible to write:

(1) $$\frac{\partial p_c}{\partial x_i} = \frac{\partial p_{nw}}{\partial x_i} - \frac{\partial p_w}{\partial x_i} ,$$

(2) $$q_{wi} = \frac{k_w}{\mu_w} \left(- \frac{\partial p_w}{\partial x_i} + \rho_w g_i\right) ,$$

(3) $q_{nwi} = \frac{k_{nw}}{\mu_{nw}} (- \frac{\partial p_{nw}}{\partial x_i} + \rho_{nw} g_i)$,

(4) $\frac{\partial q_{nwi}}{\partial x_i} = \phi \frac{\partial S}{\partial t}$.

For the special case of steady flow of two fluid phases, only the first three equations are combined to give

$$\frac{\partial p_c}{\partial x_i} = (\frac{q_{wi}\ \mu_w}{k_w} - \frac{q_{nwi}\ \mu_{nw}}{k_{nw}}) - (\Delta\rho) g \quad . \qquad 4.2$$

4.2 STEADY FLOW OF A WETTING FLUID

Equation 4.2 can be written as

$$\frac{\partial p_c}{\partial x_i} = \Delta(\frac{q_i \mu}{k}) - (\Delta\rho) g_i$$

in which Δ denotes the difference between the quantity evaluated for the wetting phase and that evaluated for the non-wetting phase.

If only the wetting phase is flowing, Δ can be dropped from the first term so that

$$\frac{\partial p_c}{\partial x_i} = \frac{q_{wi} \mu_w}{k_w} + (\Delta\rho) g \frac{\partial h}{\partial x_i} \quad . \qquad 4.3$$

Solving Equation 4.3 for q_{wi} gives

$$q_{wi} = \frac{k_w}{\mu_w} [\frac{\partial p_c}{\partial x_i} - (\Delta\rho) g \frac{\partial h}{\partial x_i}] \qquad 4.4$$

which is the flux equation for the steady flow of a wetting fluid in a two-phase fluid system.

4.2.1 Steady downward flow of water through a homogeneous petroleum reservoir - It often happens that accumulations of oil are found in lithological traps through which water is flowing steadily "downdip". The trap consists of an aquifer between impermeable strata which are inclined at some angle with respect to a horizontal plane. At some point in the aquifer a porous rock having a p_e higher than that of the remainder of the aquifer has blocked the upward migration of oil which otherwise would have occurred due to the lower density of oil compared to the brine which originally saturated the entire aquifer. A lithological trap of this type is illustrated in Figure 4-1.

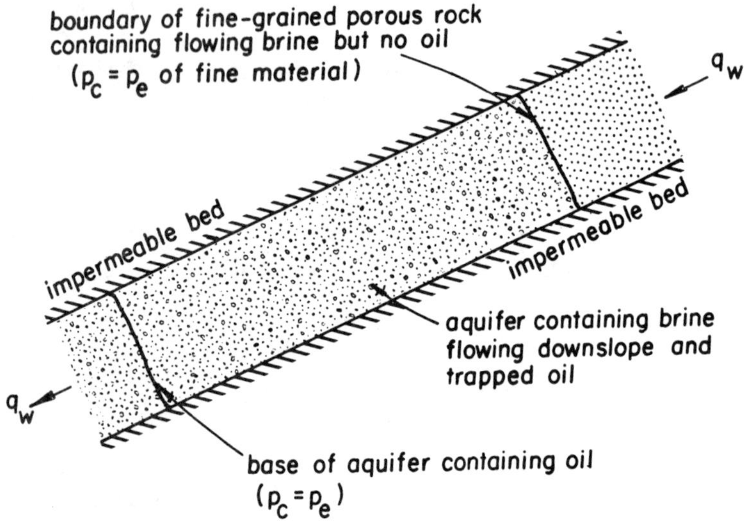

Figure 4-1. Lithological oil trap with brine flowing steadily downdip.

The questions to be answered in regard to the oil trap include:

(1) What is the extent of the oil body below the "cap" rock?
(2) How much oil is in the trap?
(3) How is the oil distributed?
(4) Where is the best place to drill wells to recover the oil?

To answer these questions, the distribution of p_c within the oil body must be determined. The maximum p_c occurs at the cap rock (oil into brine). Capillary pressure cannot be higher than p_e of the cap rock without oil leaking through the cap rock and being lost. Likewise, oil cannot exist as an interconnected phase in the aquifer below an elevation where p_c is equal to p_e of the rock containing the oil body.

The distribution of p_c within the oil body can be determined by applying these boundary conditions to a solution of Equation 4.3. It is assumed that the downdip flow is 1-dimensional and that the aquifer containing the oil body is homogeneous and has a uniform slope making an angle β with a horizontal plane. Since the flow is steady, q_{wi} is treated as a constant. However, k_w is a function of p_c.

For the case under consideration, q_{wi} is a negative component if x_i is regarded as being positive upslope, but $\partial h/\partial x_i$ is positive and is equal to $\sin \beta$. It is slightly easier to plot p_c as a function of elevation h rather than x_i. Consequently, it is noted that x_i is equal to $h/\sin \beta$ and Equation 4.3 becomes

$$\frac{dh_c}{dh} = 1 - \frac{|q_w| \mu_w}{(\Delta\rho) g\, k_w \sin \beta} ,$$

where h_c is $p_c/(\Delta\rho)g$, the capillary pressure head. Because the flow is steady and 1-dimensional, the flow equation is now an ordinary differential equation.

A solution can be obtained numerically if experimental data for $k_w(p_c)$ are available. It can be more easily solved by use of the Brooks-Corey relationship

$$k_w = k_m \left(\frac{p_d}{p_c}\right)^{\eta}$$

where η is $3\lambda + 2$. It usually happens that at the lower boundary of the interconnected oil body, where p_c is p_e, k_w is approximately $0.5\, k_m$.

The flow equation can be written in the form

$$\frac{dh_c}{dh} = 1 - a\left(\frac{h_c}{h_d}\right)^{\eta} \qquad 4.5$$

in which

$$a = \frac{|q_w| \mu_w}{(\Delta\rho) g\, k_m \sin \beta}$$

and h_d is $p_d/(\Delta\rho)g$.

The solution of the flow equation is of the form

$$\int dh = \int f(h_c)dh_c \qquad 4.6$$

in which

$$f(h_c) = (1 - a(\frac{h_c}{h_d})^{\eta})^{-1}$$

so that values of h corresponding to particular values of h_c can be obtained by integration. In particular, the elevation difference over which the oil body may extend can be obtained by integrating Equation 4.6 over a range of h_c from $p_e/(\Delta\rho)g$ of the aquifer material to $p_e/(\Delta\rho)g$ for the cap rock.

Unfortunately, the integration is usually difficult to perform analytically because η is usually greater than 6.0. In any case, the analytical expression for the integral is very complex so that it is much easier to use a computer program to integrate the function for any value of η. Such a program has been developed by P. R. Corey et al. (1982).

Some qualitative aspects of the relationship $h_c(h)$ can be determined by inspection of Equation 4.5, noting that q_w is typically a very small number. When h_c is equal to $p_e/(\Delta\rho)g$, and $k_w \cong 0.5\ k_m$, dh_c/dh is only slightly less than 1.0. Thus the distribution of h_c is practially hydrostatic at the bottom of the interconnected oil body. Furthermore, dh_c/dh theoretically cannot be negative because this would imply a decreasing h_c which restores the value of dh_c/dh to zero. Consequently, dh_c/dh is approximately 1.0 at the bottom of the interconnected oil body and approaches zero if the oil body extends over a sufficient depth.

A plot of $h_c(h)$ is shown in Figure 4-2. Note that the transition (of the $p_c(h)$ function from a practically hydrostatic relationship to a value of p_c that is essentially invariant with elevation) is abrupt. This is because η is usually large, a value of 8 being typical. The larger the value of η, the more abrupt the transition.

Consider a case in which p_e of the cap rock is twice the value of p_d for the aquifer rock and η for the aquifer material is 8. It is desired to determine the value of $|q_w|$

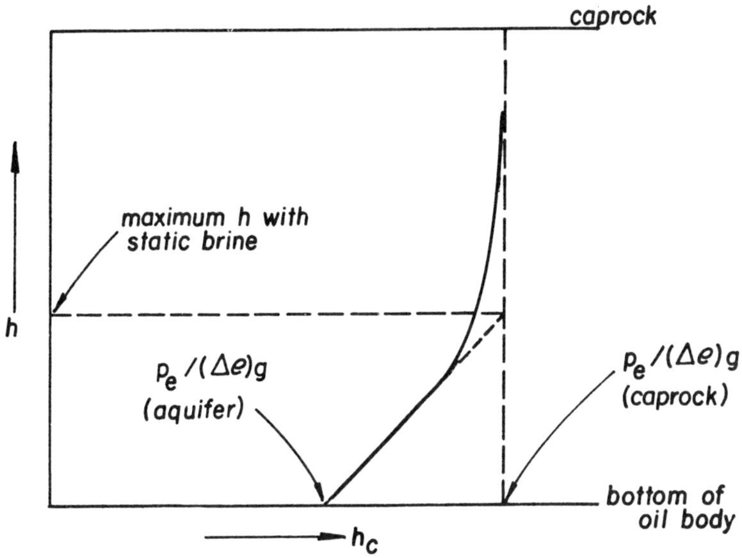

Figure 4-2. Distribution of capillary pressure with elevation in lithological oil trap with brine flowing downslope.

that will permit the oil body to reach a maximum saturation at the cap rock and the oil body to extend for an indefinite distance with this oil saturation at the cap rock.

Such a condition exists when dh_c/dh approaches zero and p_c equals p_e of the cap rock. The value of $|q_w|$ can be determined from Equation 4.5 by setting dh_c/dh equal to zero. In this case, 'a' in Equation 4.5 is 2^{-8}, and

$$|q_w| = \frac{(\Delta\rho)g\ k_m \sin\beta}{\mu_w(2^8)} \quad . \qquad 4.7$$

A smaller value of $|q_w|$ will result in oil leaking through the cap rock. In the limit as $|q_w|$ approaches zero, the distribution of p_c is entirely hydrostatic and the elevation interval occupied by interconnected oil is given by

$$\Delta h = \frac{\Delta p_e}{(\Delta\rho)g}$$

in which Δp_e is the difference in entry pressures of the cap rock and aquifer materials.

A value of $|q_w|$ larger than that given by Equation 4.7 causes the oil body to be stretched over a larger h, but the value of p_c is smaller than p_e of the cap rock and the oil saturation is smaller. When $|q_w|$ exceeds that given by

$$|q_w| = \frac{(\Delta\rho) g\, k_m \sin\beta}{\mu_m} \quad ,$$

an oil body cannot exist as an interconnected phase because dh_c/dh necessarily would be negative.

It sometimes happens that flow is actually upslope. In such cases, the maximum oil accumulation is even smaller than that given by

$$\Delta h = \frac{\Delta p_e}{(\Delta\rho) g} \quad .$$

In fact, oil bodies have not been observed where flow of brine is upslope although, surprisingly, upslope flows are not uncommon.

In cases where flow is downslope through a lithological trap, it is clear that the most appropriate place to drill oil wells is in the upper part of the oil body near the cap rock. In this region the proportion of brine produced along with the oil should be less than elsewhere in the trap.

In aquifers that are not homogeneous, distributions of p_c are different from that shown in Figure 4-2. This situation is discussed in Section 4.2.5.

4.2.2 Downward flow of water through a homogeneous soil profile to a water table - Equation 4.5 applies also to steady downward flow through a soil in which air is present as a non-wetting phase.

For this case, it is feasible to assume that both ρ_{nw} and p_{nw} are equal to zero, and $\sin\beta$ is 1.0, so that

$$\frac{dh_c}{dh} = 1 - \frac{|q_w|}{K_m} \left(\frac{h_c}{h_d}\right)^{\eta}$$

for $h_c \geqq h_d$. It is often convenient to scale the variables h_c and h by dividing each by h_d. The ratio $|q_w|/K_m$ is also regarded as a scaled variable designated as $\hat{q}$. In terms of scaled variables the flow equation becomes

$$\frac{d\hat{h}_c}{d\hat{h}} = 1 - \hat{q}\,\hat{h}_c^{\eta} , \qquad 4.8$$

which, however, assumes that $\hat{h}_c$ is greater than 1.0. Again, a solution can be obtained by numerical integration [P. R. Corey et al. (1982)].

When flow is steadily downward to a water table, the result is qualitatively like that shown in Figure 4-2 for downward flow through an oil trap. In this case, however, there is no cap rock to set a limit on the value of $\hat{h}_c$. Instead, the source of air at atmospheric pressure at the soil surface means that the air pressure will never become less than atmospheric. Theoretically, there is no limit to the extent that the water pressure can be lowered by reducing $\hat{q}$ and increasing $\hat{h}$.

When $\hat{q}$ becomes greater than 1.0, $d\hat{h}_c/d\hat{h}$ becomes negative and an interconnected air phase does not exist. In the limit, as $\hat{q}$ approaches zero, the distribution of $\hat{h}_c$ becomes everywhere hydrostatic. Figure 4-3 shows a family of distributions for a range of values of $\hat{q}$.

In Figure 4-3, $\hat{q}_1$ is less than 1.0 so that $\hat{h}_c$ is not negative, but it is large enough so that p_c is everywhere less than p_e, and an interconnected air phase does not exist. When this situation exists, Equation 4.8 does not apply. Instead,

$$\frac{d\hat{h}_c}{d\hat{h}} = 1 - \hat{q} \qquad 4.9$$

so that $d\hat{h}_c/d\hat{h}$ is a constant. In any case, Equation 4.9 must be used for all values of $\hat{h}_c$ less than p_e/p_d, a ratio slightly greater than 1.0.

Another way of arriving at the same result for field situations is by redefining K_m as the value of K existing at p_e on whatever cycle (wetting or drying) which exists in the field situation. Also, $\hat{h}_c$ and $\hat{h}$ are scaled in terms of p_e rather than p_d. In this case, Equation 4.8 applies for $\hat{h}_c \geqq 1.0$ and Equation 4.9 for $\hat{h}_c \leqq 1.0$. If the problem under consideration is one that occurs on a wetting cycle, the value

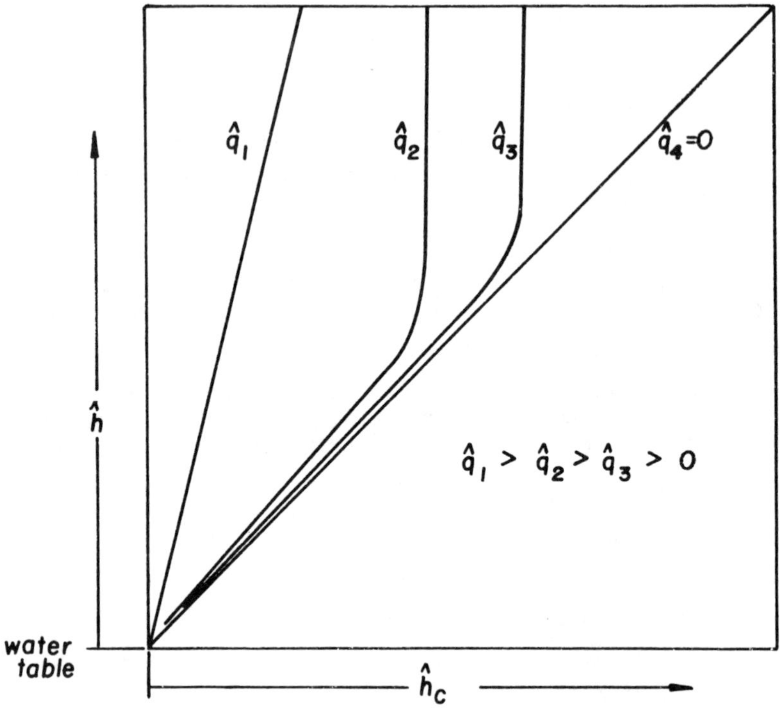

Figure 4-3. Capillary pressure head as a function of elevation during steady downward flow of water through a homogeneous soil.

of p_e is the value of p_c existing at the boundary below which the air phase is no longer interconnected.

The behavior indicated in Figure 4-3 makes possible the long-column method of determining $k_{rw}(p_c)$. Steady downward flow is established at a range of values of $\hat{q}$ such that the value of $\hat{h}_c$ is practically constant with $\hat{h}$ near the upper part of the column. The value of $\hat{h}_c$ may be determined with a tensiometer at that part of the column, and the value of S can be determined using gamma attenuation. Since the value of pressure is constant in this region of the column, the value of $\partial H/\partial z$ is 1.0 and K_w is equal to $|q_w|$. The procedure will not give satisfactory results, however, unless the column is homogeneously packed and is sufficiently long that p_c is actually uniform at the observation point.

The length of column needed depends upon the p_e of the soil, since $d\hat{h}_c/d\hat{h}$ is a constant equal to $1 - \hat{q}$ for values of $\hat{h}_c < 1.0$. Where $\hat{h}_c > 1.0$, the value of the product $\hat{q}\,\hat{h}_c^{\eta}$ increases toward a value 1.0. The transition is short if η is large and somewhat longer if η is relatively small. Consequently, the length of the column needed depends upon η as well as p_e, but the latter is usually the more important factor.

A complication may arise when $K_w(p_c)$ is determined on the drainage cycle, that is, when $\hat{q}$ is decreased in increments starting with a fully saturated column. When $\hat{q}$ is reduced, a period of time is required before a new steady state exists. Although it is possible for $\hat{q}$ to be decreased immediately throughout the column, it is not possible for air to immediately replace the water which is necessary for a new steady state to exist. During this period the value of k_w is higher than the steady state value. Consequently, $\hat{q}$ is smaller and $d\hat{h}_c/d\hat{h}$ is larger than desired. The top of the column may desaturate too much. A steady state is eventually reached, but it may be reached from the wetting direction rather than the drainage direction intended. This will produce erratic data for $K_w(p_c)$, but $K_w(S)$ is usually not affected significantly.

The problem of reversing cycles can be minimized by making the increments of q_w small and providing sufficient air vents along the length of the column.

4.2.3 Steady downward flow to a high p_c sink - The length of column needed to produce a practically uniform p_c near the top of a column can be substantially reduced by maintaining a high value of p_c at the bottom of a column of homogeneous soil.

In this case Equation 4.8,

$$\frac{d\hat{h}_c}{d\hat{h}} = 1 - \hat{q}\,\hat{h}_c^{\eta} ,$$

applies to the entire column provided $\hat{q} < 1.0$. This is because $h_c > p_e/(\Delta\rho)g$ everywhere in the column. Again, the solution is

$$\int f(\hat{h}_c)\, d\hat{h}_c = \int d\hat{h}$$

which can be integrated numerically using a computer. The integration begins at the lower boundary where $\hat{h}_c$ is at a known value greater than 1.0, say 3.0. The integral gives the value of $\hat{h}$ at which a particular value of $\hat{h}_c$ occurs.

A curve obtained by this procedure is shown in Figure 4-4 as the solid line. This is compared to an analogous curve (for the same $\hat{q}$ and η) obtained for downward flow to a water table. Note that for steady downward flow to both a high p_c sink and a water table, the value of $\hat{h}_c$ in the upper part of the column approaches the same asymptote. The value of $\hat{h}_c$ at the asymptote depends on q and η. This is the value of $\hat{h}_c$ for which $d\hat{h}_c/d\hat{h}$ is zero. With a high p_c sink, however, the asymptote is approached over a much shorter length of column.

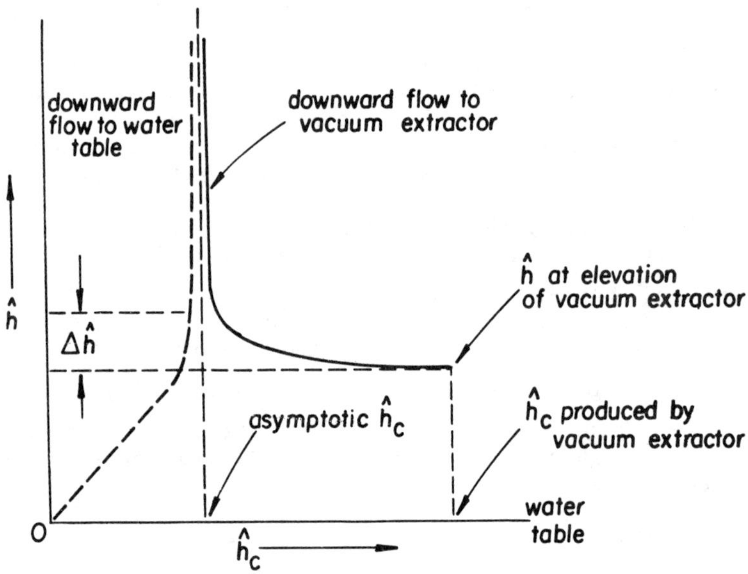

Figure 4-4. Distribution of $\hat{h}_c$ during steady downward flow to a high p_c sink compared to flow to a water table. [Adapted from P. R. Corey et al. (1982)].

The behavior illustrated in Figure 4-4 permits the determination of $K_w(p_c)$ or $K_w(S)$ using columns shorter than those needed for the long-column procedure. There is an important difficulty, however, in that high p_c sinks are not easy to maintain. It would seem that the only requirement would be a capillary barrier at the base of a column connected to a siphon tube operated at the necessary vacuum.

Unfortunately, the resistance of a capillary barrier (and especially the contact zone between barrier and the soil column) has a tendency to increase rapidly during operation. What starts out to be a high p_c sink may quickly become a barrier to flow, thus invalidating the experiment. Experience has shown that a capillary barrier consisting of a sequence of unconsolidated granular layers, coarser at the bottom and progressively finer at the top in contact with the soil, is the most satisfactory arrangement. Such a barrier has much less tendency to become plugged or to lose contact with the soil than a rigid or semi-rigid porous membrane. The p_c of the finest layer in the barrier should not be larger than necessary to hold the vacuum of the outflow tube. Also, the fine layer should not contain clay.

Another interesting application of a high p_c sink is for monitoring the soil water solution passing through a soil profile. There are numerous occasions for which a measurement of the quantity and quality of the soil leachate is needed. For example, environmental engineers employ such methods for studying pollution of groundwater from a variety of sources.

The principle problem to be overcome is due to the soil water being at some pressure less than atmospheric. Therefore, to get the water into a collection device, the pressure of water in the device (in contact with soil water) must be lower than the soil water pressure. This must be done without permitting the device to fill with air, thus excluding the soil water. However, the soil water pressure at the entrance to the collector must not be greatly lower than that of the undisturbed soil water because this will cause excessive convergence of streamlines and invalidate a quantitative determination of the rate at which leachate passes a particular horizontal plane.

Several investigators have utilized porous ceramic capillary barriers operating under a suction for this purpose. Cole (1968) used this technique to measure rate of water flow through a forest soil. Convergence of streamlines towards the capillary barrier was prevented by manual adjustment of the suction to that of the water in the surrounding soil.

Duke, Kruse and Hutchinson (1970) improved on Cole's device by installing an automatic system to adjust the suction in the collector to that in the surrounding soil. It was found, however, that the automatic equipment was difficult to maintain as well as being expensive to install. Duke and McWhorter noticed that when suction is applied at the bottom of a buried trough (as shown in Figure 4-5) the amount of leachate collected is insensitive to the amount of suction applied. P. R. Corey et al. (1982) investigated the question of how deep a trough would be required to ensure that the suction at the trough would depend only on the vertical component of $\underline{q}_w$ and would not be affected significantly by the suction at the bottom of the trough.

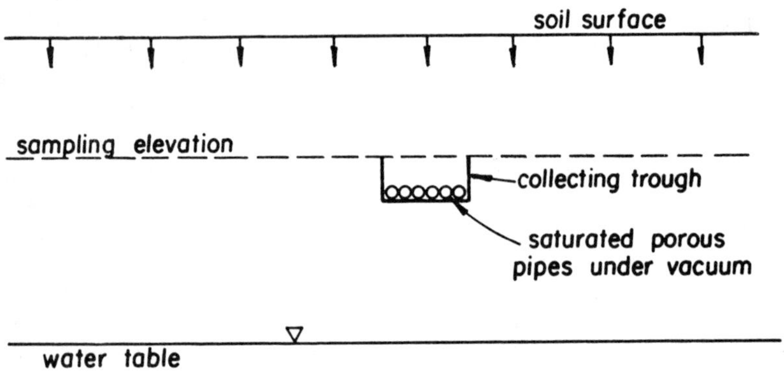

Figure 4-5. Sketch of soil water monitoring system.

This problem is tantamount to that of determining the magnitude of $\Delta\hat{h}$ (indicated in Figure 4-4) such that $\hat{h}_c$ is within some arbitrarily small increment of the asymptotic value. When $\hat{h}_c$ is at the asymptotic value, no convergence of flow lines toward the top of the trough should occur. The magnitude of $\Delta\hat{h}$ is, of course, greater for soils having small η and for smaller values of $\hat{q}$. It also is affected slightly by the magnitude of $\hat{h}_c$ applied at the bottom of the trough.

It was found that a trough 20 cm deep would be satisfactory for typical soil conditions but might not be satisfactory for a soil having a very wide range of pore sizes, say η less than about 6.

4.2.4 Steady upward flow from a water table - There is general agreement that the rate of evaporation from a fallow soil may be controlled by either the capacity of the atmospheric environment to evaporate water or the capacity of the soil to transmit water to the surface. Except where the water table is at very shallow depths, the capacity of the soil to transmit water usually is the limiting factor [Anat et al. (1965)].

When upward flow is from a stationary water table, the flux rate may sometimes approach a steady state, and the limiting rate may be determined by application of Equation 4.3. In this case, Equation 4.3 is written as

$$\frac{d\hat{h}_c}{d\hat{h}} = 1 + \hat{q} \quad \text{for} \quad \hat{h}_c \leqq 1$$

and

$$\frac{d\hat{h}_c}{d\hat{h}} = 1 + \hat{q}\,\hat{h}_c^{\eta} \quad \text{for} \quad \hat{h}_c > 1 \quad . \qquad 4.10$$

Note that Equation 4.10 differs from Equation 4.8 in that $\hat{q}$ is positive rather than negative. The value of $d\hat{h}_c/d\hat{h}$, therefore, cannot equal zero anywhere in the soil profile and has a minimum value of 1.0. Figure 4-6 presents a typical solution of Equations 4.10 giving the variation of $\hat{h}_c$ with $\hat{h}$.

Because the change in pressure due to flow and that due to gravity are additive, the transition from wet to dry soil is very abrupt. For any particular soil (and distance from dry soil to water table) there is a maximum upward flux rate. This fact was first pointed out by Gardner (1958). The maximum upward flow rate (for a given depth from dry soil to water table) occurs when the dry soil is encountered only at the extreme surface, or perhaps when the surface is not completely dry. Faster drying causes the upward flow rate to decrease, evidently because of contact angle hysteresis or some such mechanism. The latter phenomenon has been discussed by Schleusener and Corey (1959).

Equation 4.10 can be written as

$$\int f(\hat{h}_c)\, d\hat{h}_c = \int d\hat{h}$$

and integrated numerically. The integration is carried out in two steps. The first step is $\hat{h}_c$ from zero to 1, for which $f(\hat{h}_c)$ is a constant equal to $1/(1 + \hat{q})$ and the second step is $\hat{h}_c$ from 1 to ∞, the value of h_c for the dry soil being assumed to approach infinity. The integration is performed by substituting arbitrary values of $\hat{q}$ and η. The sum of the

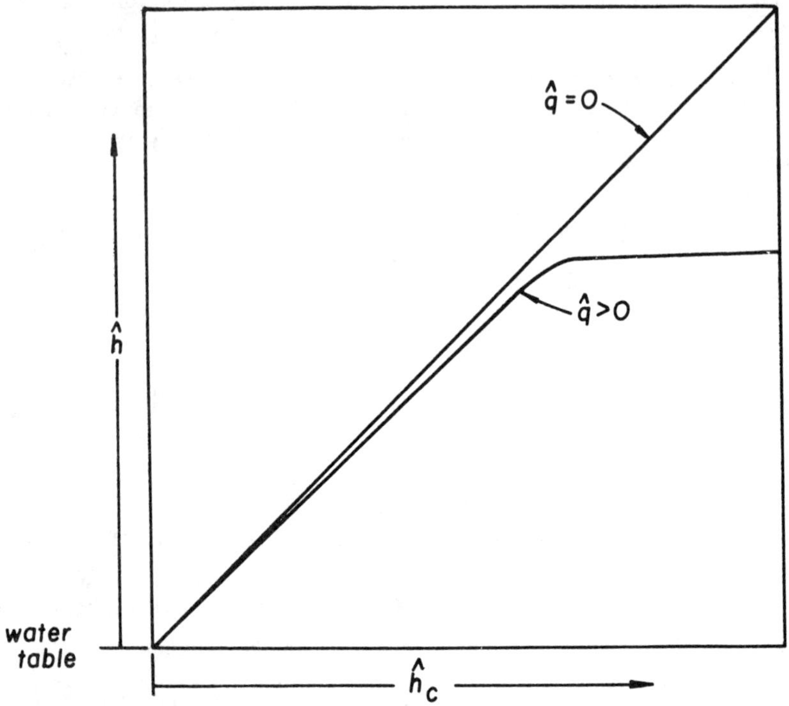

Figure 4-6. Distribution of capillary pressure head during steady upward flow from a water table.

two integrals gives the scaled depth $\hat{d}$ from dry soil to water table corresponding to the value of $\hat{q}$ and η selected. By repeating the process for a range of values of $\hat{q}$ and η, H. R. Duke obtained a nomograph [Anat et al. (1965)] showing the relationship among $\hat{q}$, $\hat{d}$ and η.

Anat developed an approximate analytical expression for $\hat{q}$ as a function of $\hat{d}$ and η. He did this by expanding $f(\hat{h}_c)$ into a convergent series and integrating term by term. For values of $\hat{q} < 0.01$ (which includes the range of most practical applications), the relationship is

$$\hat{q} \approx \left(1 + \frac{1.886}{\eta^2+1}\right)^{\eta} \hat{d}^{-\eta} \quad . \qquad 4.11$$

Equation 4.11 is somewhat comparable in form to equations presented earlier by Gardner (1958) for integer values of η

from 1 to 4. In deriving his equations, Gardner also assumed $\hat{q}$ was small.

4.2.5 Steady downward flow through stratified media - A case is considered in which water is percolating steadily downward through a sequence of layers to a deep water table. The layers have contrasting properties but are homogeneous within themselves. For simplicity, a system is considered in which alternating layers of only two types of media exist. The $K_w(p_c)$ relationships for these two media are shown in Figure 4-7 on log-log plots. The arrangement of the layers is as shown in Figure 4-8.

Downward flow occurs at a steady flux rate which is indicated in Figure 4-7. Continuity requires that the flux rate be the same in all layers since the flow is steady. Furthermore, the water is interconnected (continuous) throughout all layers, including at the boundaries between layers, otherwise there could be no flow. The water pressure, therefore, is continuous at all points in the column because a pressure discontinuity in a continuous liquid phase would imply an infinite driving force at the point of discontinuity in pressure.

The value of dh_c/dh at all points is given by Equation 4.3 which for the purpose of this analysis is written as

$$\frac{dh_c}{dh} = 1 - \frac{|q_w|}{K_w}$$

in which K_w is a function of h_c. The solution of the equation is

$$\int f(h_c)dh_c = \Delta h \quad .$$

When the integration is carried out across boundaries between layers, however, there is an abrupt change in K_w which produces a discontinuity in dh_c/dh. Consequently, h_c is a continuous function of the space coordinates but not an analytic function in this case.

The shape of the curve shown in Figure 4-8 can be understood by reference to the differential equation of flow. Note that when K_w equals $|q_w|$, dh_c/dh is zero so that the hydraulic gradient is 1.0 and the driving force is entirely gravitational. This situation exists in the regions indicated by the vertical portions of the curve.

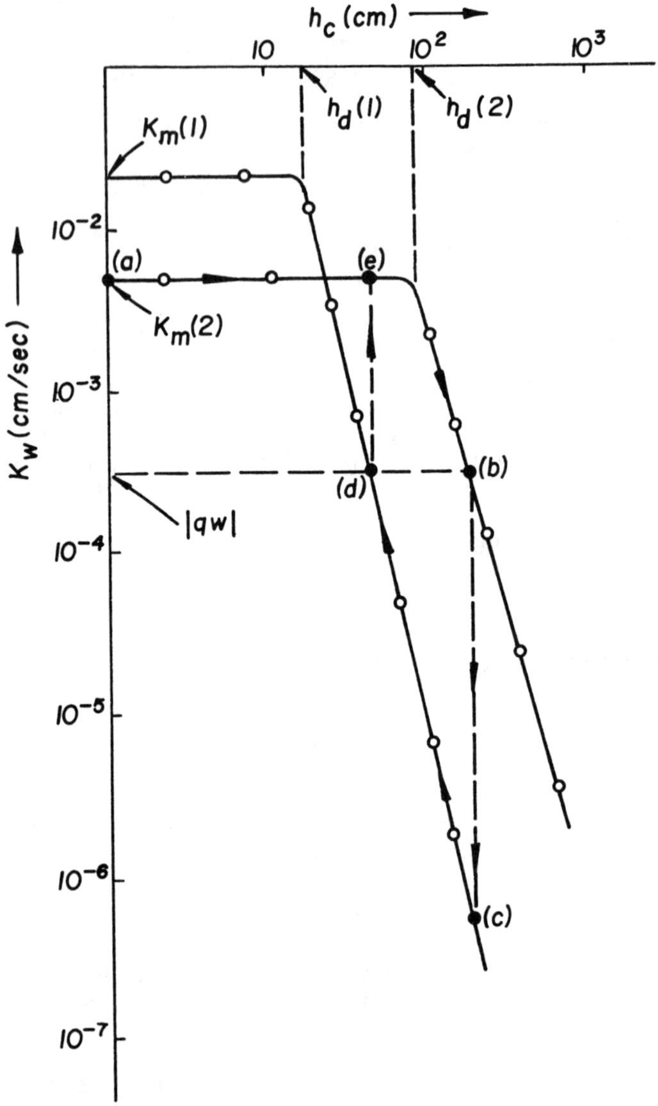

Figure 4-7. K_w as a function of p_c (for two media from a layered soil) on a log-log plot.

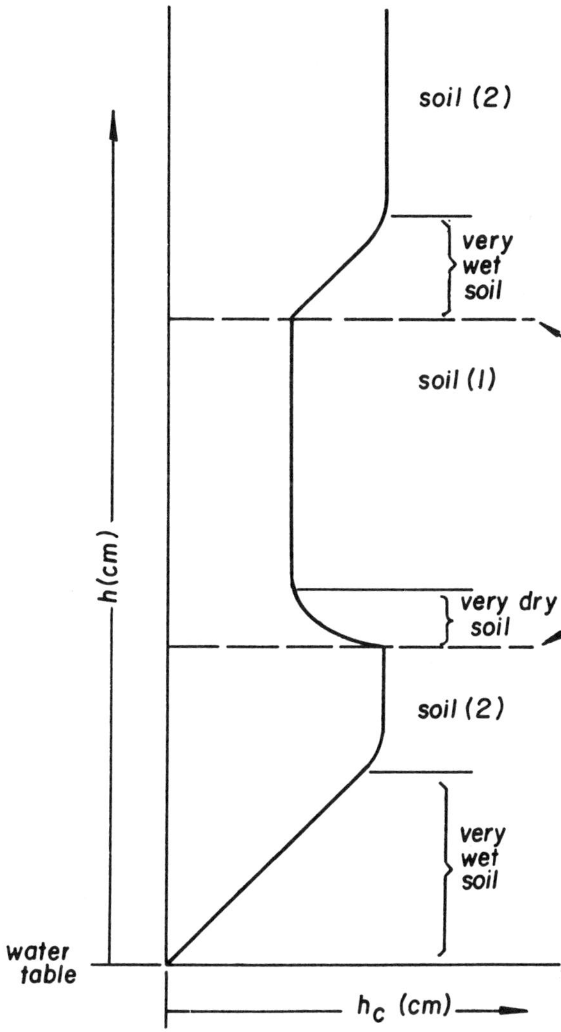

Figure 4-8. Distribution of capillary pressure during steady downward flow of water through layered soils.

It is informative to consider the behavior of dh_c/dh starting at the water table in soil (2). The point (a) on the $K_w(h_c)$ curve for soil (2) indicates the value of h_c and K_w at this point. Immediately above this point, h_c increases practically hydrostatically because $|q_w| << K_m$ (2). This trend continues until h_c exceeds h_d (2). The curve then bends until h_c reaches a value indicated by point (b) in Figure 4-7; where $K_w = |q_w|$. The curve is then vertical because dh_c/dh is zero. Where soil (1) is encountered, however, the value of K_w is as indicated by point (c). This value is extremely small. The ratio $|q_w|/K_w$ is large, and dh_c/dh has a large negative value. This causes h_c to decrease rapidly to a value indicated by point (d) where K_w again is equal to $|q_w|$ so that the curve is again practically vertical. At this value of h_c, however, K_w in soil (2) is as indicated by point (e). Where this soil is encountered, the pressure distribution is practically hydrostatic until h_c (indicated at point (b)) is again reached and the curve again becomes vertical.

As indicated in Figure 4-8, the wettest soil (other than near the water table) is just above the coarse material (soil 1). Also, the driest soil is in the coarse material just above the underlying fine soil. In fact, the resistance to flow in the thin dry layer on top of the fine soil, is probably greater than the combined resistance of the remainder of the profile. In other words, the flux rate $|q_w|$ is controlled by this region of the profile. Conditions analogous to this case occur frequently in nature and for this reason, layered soils drain much more slowly than homogeneous soils.

The behavior illustrated in Figure 4-8 was demonstrated experimentally by Scott and Corey (1961) who were the first to apply Equation 4.3 to flow through layered soils. They showed that Equation 4.3 described the experimentally measured pressure distributions in layered soils precisely for the cases they investigated.

Equation 4.3, in its more general form, also can be used to describe the pressure distribution during steady flow of brine through a lithological oil trap in which the aquifer material varies in texture along the path of flow. The relationships illustrated in Figure 4-8 hold for this case also. Therefore, it can be predicted that oil concentrations are highest in regions of coarse sands just upslope from finer-textured materials. The highest brine saturations are in the finer materials immediately upslope from coarser materials.

4.3 STEADY FLOW OF A NON-WETTING FLUID

Steady flow of a non-wetting fluid rarely occurs in field situations. Consequently, this subject is of concern mainly in respect to laboratory experiments. Two such experiments are discussed in Chapter III. One involves the simultaneous flow of two fluid phases under the same pressure gradient in order to obtain uniform saturations for the measurement of k_w and k_{nw} as functions of S. The other involves steady upward flow of air only, under a pressure gradient equal to the static pressure gradient in the liquid wetting phase. This is done in order to obtain $k_{nw}(S)$ on porous rock cores.

Another experiment, considered here, is an academic exercise to illustrate the physics of non-wetting phase flow. This involves horizontal flow of a non-wetting fluid in the presence of a static wetting phase.

A cylindrical core of porous rock of small diameter is considered. The surface of the cylinder is sealed except for the ends. The porous cylinder is first fully saturated with a liquid. A pressure difference is imposed in the air across the two ends of the core which is sufficient to displace some of the liquid and to establish a flow of air through the medium. Assuming that the air is presaturated with the vapor of the liquid, the flow will eventually reach a steady state and the liquid retained in the porous rock will become static. The situation is illustrated in Figure 4-9.

The version of Equation 4.3 applicable after the liquid becomes static is

$$\frac{dp_c}{dx} = - \frac{q_a \mu_a}{k_a}$$

where the subscript "a" refers to air. In order for air to pass entirely through the core, p_a must exceed p_ℓ by an amount at least equal to p_e for the particular medium and fluids considered. The subscript "ℓ" refers to liquid.

For this problem, it is assumed that p_c at the outflow face is just equal to p_e. Since the liquid is static, p_ℓ is a constant along the line of flow. It is convenient to use p_ℓ as a datum from which to measure the pressure of the air

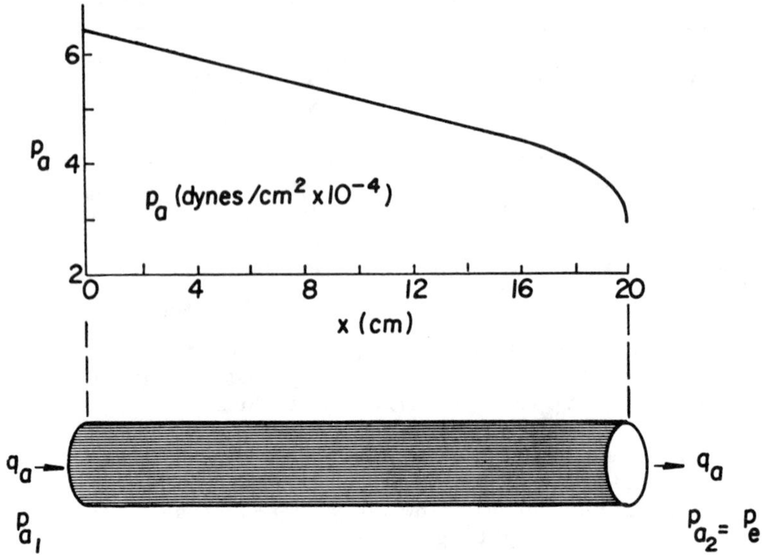

Figure 4-9. Horizontal porous cylinder containing static liquid and flowing air.

because p_c is then equal to p_a and k_a is a function of p_a.

The air pressure at any point x along the centerline of the cylinder is given by

$$\int_{p_{a_1}}^{p_a} k_a dp_a = - q_a \mu_a x$$

The integral on the left side of this equation can be evaluated if $k_a(p_c)$ is known. One way of doing this is to employ Equation 3.54, that is,

$$k_a = k_m \{[1 - (\frac{p_d}{p_a})^\lambda]^2 [1 - (\frac{p_d}{p_a})^{2+\lambda}]\} \ .$$

In any case, the integration can be performed numerically using a computer.

The pressure distribution shown in Figure 4-9 was determined in this way, for a case in which $\lambda = 2$. Note that much of the pressure drop occurs near the outflow face. This is because the liquid retained in the core tends to accumulate at the outflow face and resistance to flow at this face is large relative to that through other parts of the core.

4.4 STEADY FLOW TOWARD PARALLEL DRAINS

The problem considered here is to a large extent hypothetical although it has considerable practical importance. In many parts of the world, agricultural drains are designed (in respect to depth and spacing) upon the assumption that water percolates steadily downward to the water table. Upon reaching the water table, the water flows laterally to some sink. In order to maintain the water table at a safe depth below the surface, artificial sinks may be provided in the form of parallel drains at some depth below the surface, preferably near an impermeable stratum if such a stratum exists.

The hypothetical model usually used as a basis of design is illustrated in Figure 4-10. This model is idealized in several respects. First, the recharge rate w is assumed to be constant in respect to both time and areal distribution. Secondly, the drains are pictured as open ditches penetrating to some constant depth, whereas in reality, the drains are usually buried conduits of relatively small size, say 15 to 20 cm tile. The water table is regarded as being in equilibrium under the existing conditions of recharge and drainage. It is also assumed that the Dupuit approximations are valid for this case and that the aquifer is homogeneous and isotropic.

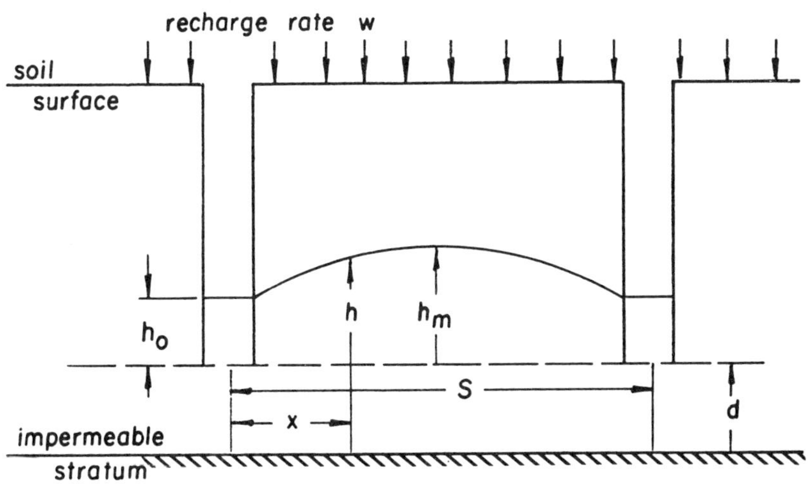

Figure 4-10. Flow toward parallel drains with uniform recharge.

In the usual analysis of this problem, it is also assumed that the water table is the upper streamline. This latter assumption may often lead to serious error, and it is the purpose of this section to show how this assumption can be avoided. However, the standard analysis is presented first.

The origin of coordinates is taken on the impermeable layer below the center of the drain on the left. A vertical plane drawn between the center of the two drains at $x = S/2$ is regarded as a divide across which no flow takes place. To the left of the central divide, all flow eventually reaches the left drain, and to the right of the divide flow is toward the right drain.

Flow through a vertical plane at x is considered. The flow through this section is toward the left drain and is equal to the recharge on the surface between the section and the divide. The surface area on which this recharge takes place (for a slice of aquifer of unit thickness) is equal to $S/2 - x$. The quantity of water flowing per unit time through the vertical plane of unit width is, therefore,

$$Q_x = w(\frac{S}{2} - x) \quad .$$

According to the Dupuit approximation, Q_x should also be given by

$$Q_x = Kh \frac{dh}{dx}$$

where h is the elevation of the water table above the impermeable layer. Equating the two expressions for Q_x gives

$$(\frac{S}{2} - x)w = Kh \frac{dh}{dx} \quad .$$

Separating the variables and integrating results in

$$(\frac{wS}{2})x - (\frac{w}{2})x^2 = \frac{K}{2} (h^2 - h_o^2) \quad . \qquad 4.12$$

By transforming the origin to the midpoint between drains, this can be written in the form

$$\frac{h^2}{c_1} + \frac{x^2}{c_2} = 1$$

which is the equation of an ellipse.

In the usual case $h_o << h_m$ so that h_o can be considered as equal to zero. In this case, the flow equation can be integrated between the limits $(x = 0, h = d)$ and $(x = S/2, h = h_m + d)$ which gives

$$S^2 = \frac{4Kh_m}{w}(h_m + 2d) \quad . \qquad 4.13$$

Equation 4.13 has been used in many parts of the world and especially by the U.S. Soil Conservation Service for determining the spacing of relief drains. The equation is sometimes known as the ellipse equation. In other places it is called the Hooghoudt, the Donnan, or the steady state equation. It is more appropriate for areas of frequent rainfall, such as Holland where it originated, than for irrigated regions. In such cases, the assumption of a constant recharge and an equilibrium state of the water table may be a reasonable approximation.

If the height d is not relatively small compared to h_m, the Dupuit approximation will fail rather seriously because the flow into the drain will approach radial rather than horizontal flow. In the latter case, a different equation for the drain spacing is necessary. For example, according to Equation 4.13, as d becomes infinite, S also becomes infinite. This is not realistic, the difficulty being that the equation fails to account properly for radial flow in the vicinity of the drain.

Equation 4.13 also fails for cases in which the so-called capillary fringe $(p_d/\rho_w g)$ designated by h_d is not small compared to h_m. This is a frequent situation, but until recently little attention has been given to accounting for the flow in the region above the water table. The following analysis was suggested by Duke (1973) as a way of accounting for this flow.

The effectiveness of the region above the water table in conducting water horizontally toward a drain is given by an expression suggested by Myers and van Bavel [Bouwer (1964)] as

$$H_k = \frac{1}{K_m} \int_o^H K_w dh \qquad 4.14$$

where H_k is the effective permeable height, a hypothetical height of saturated soil having the same capacity to transmit water as does the partially saturated region above the water table. The maximum field conductivity is designated by K_m and

K_w is the conductivity at any other saturation. The elevation above the water table is designated as h, and H is the elevation to the soil surface.

When $H < h_e$, the soil remains saturated to the surface and $H_k = H$, h_e being the entry pressure head $p_e/\rho_w g$. If $H > h_e$, K_w is dependent upon h and any vertical flux that may exist. During steady percolation to a water table, h_c is less than h at every point. Figure 4-3 illustrates this fact.

The integration indicated in 4.14 can best be performed in two steps:

$$H_k = \frac{1}{K_m} \int_o^{h_e} K_w dh + \frac{1}{K_m} \int_{h_e}^{h_s} K_w dh \quad . \qquad 4.15$$

The limit h_s refers to the value of h_c at the soil surface. In the usual case, the surface is dry and $h_s \to \infty$, the value of the integrand at this limit being equal to zero.

In order to evaluate the integrals, however, the variable of integration h must be expressed in terms of h_c. This is accomplished by application of Equation 4.3 in the form

$$\frac{dh_c}{dh} = 1 - \frac{|q_w|}{K_w}$$

so that

$$dh = \frac{dh_c}{1 - \frac{|q_w|}{K_w}} \quad .$$

For the first term on the right in Equation 4.15, K_w is K_m, a constant. Therefore, this term is

$$\frac{h_e}{1-\hat{q}}$$

in which $\hat{q}$ is the scaled flux $|q_w|/K_m$. For the second term

$$K_w = K_m\left(\frac{h_e}{h_c}\right)^{\eta}$$

so that

$$H_k = \frac{h_e}{1-\hat{q}} + \int_{h_e}^{h_s} \frac{dh_c}{\left(\frac{h_c}{h_e}\right)^{\eta}\left[1-\hat{q}\left(\frac{h_c}{h_e}\right)^{\eta}\right]} \quad . \tag{4.16}$$

Equation 4.16 can be solved easily by numerical integration (trapezoidal or Simpson's rule) using a computer. It is sufficient to assume that H_k is a constant at all values of x (see Figure 4-10). The reason is that the integrand in Equation 4.16 becomes very small when h_c is substantially larger than h_e, say about 2 or 3 times h_e. Consequently, the greater depth to the water table near the drains has negligible effect on the value of H_k.

For the case of steady flow to parallel drains, with a steady percolation rate, the variable h in the ellipse equation (in particular, h_m) can be replaced with $h + H_k$. This procedure was used by Duke to calculate the water table profile in a laboratory drainage model consisting of a soil flume. The results are shown in Figure 4-11. In this figure, the measured water table elevations are shown along with those calculated by numerical solution of Equation 4.12 with h replaced by $h + H_k$. Note that this procedure provides a much better representation of the measured data than is obtained from Equation 4.12 in its original form.

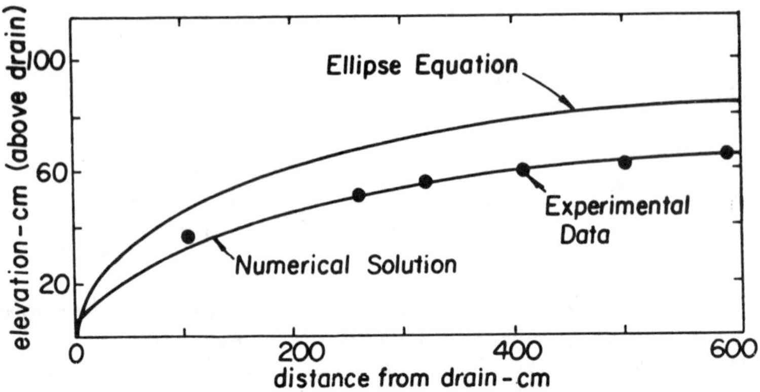

Figure 4-11. Numerical solution of Equation 4.12 corrected with Equation 4.16. [Adapted from Duke (1973)].

REFERENCES

Anat, A., Duke, H. R., and Corey, A. T. (1965). Steady upward flow from water tables. Colorado State University Hydrology Paper No. 7, June, 34 p.

Bouwer, H. (1964). Unsaturated flow in groundwater hydraulics. J. of Hydr. Div., ASCE, HY5:121-144, pp. 121-144.

Cole, D. W. (1968). A system for measuring conductivity, acidity and rate of water flow in a forest soil. Water Resources Research, Vol. 4, pp. 1127-1136.

Corey, P. R., Duke, H. R., and Corey, A. T. (1982). Monitoring soil water leachate. Trans. ASAE, Vol. 25, No. 1, pp. 96-99 and 104.

Duke, H. R. (1973). Drainage design based upon aeration. Colorado State University Hydrology Paper No. 61, June, 59 p.

Duke, H. R., Kruse, E. G., and Hutchinson, G. L. (1970). An automatic vacuum lysimeter for monitoring percolation rates. United State Department of Agriculture, ARS41-165, September, 12 p.

Gardner, W. R. (1958). Some steady-state solutions of unsaturated moisture flow equation with application to evaporation from a water table. Soil Science, Vol. 85, pp. 228-232.

Schleusener, R. A. and Corey, A. T. (1959). The role of hysteresis in reducing evaporation from soils in contact with a water table. Journal of Geophysical Research, Vol. 64, No. 4, April, pp. 469-475.

Scott, V. H. and Corey, A. T. (1961). Pressure distribution during steady flow in unsaturated sands. SSSA Proceedings, Vol. 35, No. 4, July-August, pp. 270-273.

PROBLEMS AND STUDY QUESTIONS

1. Equation 4.4 does not imply a summation over 3 orthogonal directions. However, Equation 4.1 necessarily implies a summation over 3 orthogonal directions. Explain why this is necessary, physically.

2. In the process of geophysical exploration for oil, the geophysicists look for structural "highs"; that is, they check to see if certain rock strata are encountered at higher elevations than the known elevation of the corresponding strata in surrounding areas. Explain.

3. In a confined aquifer containing both oil and brine, in which the aquifer texture is at places relatively fine-grained and in other places coarse-grained, wells that are drilled into the fine-grained material may produce mostly brine. Explain.

4. Consider a soil profile in which there is a junction between a coarse and a finer-textured material. Assume that both the coarse and the finer material are partially desaturated. Compare (1) the capillary pressures, (2) the saturations, and (3) the values of K_w of the two materials at the junction. Explain.

5. Given that a sand has a $p_e/\rho g$ of about 30 cm, estimate the length of column necessary to obtain $k_w(p_c)$ by the long-column method. How would the necessary length of column be affected if the soil under investigation is highly structured (with a low value of η being expected)? Explain.

6. When using a long column to determine $k_w(p_c)$, there is a danger that upon decreasing the flow rate by some increment, the column may over-desaturate at the top and later resaturate to some degree. Explain. This tendency is greater when columns are not supplied with sufficient air vents. Explain.

7. The tendency indicated in question number 6 may not affect the results significantly if the method is used to obtain $k_w(s)$. Explain.

8. The length of column needed to obtain $k_w(p_c)$ by the short-column method is affected by η, but not by $p_e/\rho g$. Explain.

9. Consider a case of steady downward flow of water in homogeneous sand in a long column to a water table below. If the value of q_w is 10^{-8} cm/sec and K_m is 10^{-4} cm/sec, what is K_{rw} in the upper portion of the column?

10. Consider a case of water moving steadily downward through a very long column of homogeneous sand (λ=2 and S_r=0.2) which is partially saturated. The upper half of the column has a diameter of 3 cm and the lower half has a diameter of 6 cm. The value of q for the upper portion is 1.2 x 10^{-3} cm/sec and S_e = 0.8. Ignore the situation existing near the junction of the upper and lower half and estimate the following:

(1) The value of k_w of the soil in the lower half in square microns.
(2) The value of S for the lower half.
(3) The value of k_{rw} for the upper and the lower portion.
(4) The ratio of the p_c in the upper half to that in the lower half.

11. When using a soil water extractor to determine the rate at which leachate passes a horizontal plane in a soil profile, it is important that streamlines do not converge or diverge toward the extracting device. If the extractor consists of an interceptor operating at atmospheric pressure, would you expect the streamlines to diverge or converge toward the device under usual soil water conditions? Explain.

12. Evapotranspiration rates may fluctuate greatly from hour to hour depending upon atmospheric conditions, whereas evaporation from a nearby fallow soil may not fluctuate at all during the same period of time. Explain.

13. Explain why layered soils typically drain much more slowly than homogeneous soils. Describe conditions under which a coarse stratum may control the percolation rate, and other conditions in which a fine-textured stratum may be the major restriction.

14. The vertical distribution of soil water obtained with a trickle irrigation system is likely to be much more favorable than that obtained by a sub-irrigation system. Explain fully.

15. Would you expect the effective permeable height H_k to be greater or less if the recharge rate w is increased, other conditions being equal?

16. Consider a horizontal cylinder (of small diameter) of homogeneous porous rock. The medium is initially saturated with water and is open to the atmosphere. Water is maintained at atmospheric pressure at the inflow end of the cylinder. The water pressure at the outflow end is maintained at a very low value ($p_c \to \infty$) and the outflow q approaches a maximum q_m.

Assuming that $p_c \leqq p_d$, $K_w = K_m$, and for $p_c \geqq p_d$,

$$K_w = K_m (p_d/p_c)^\eta ,$$

show that

$$q_m = K_m \frac{\eta}{\eta - 1} \frac{p_d}{\rho g L}$$

in which L is the length of the cylinder of porous rock.

17. In reference to problem number 16, assume that the rock has a very uniform pore size and $p_d/\rho g$ is twice the length of the porous cylinder. Estimate k_m in square centimeters if q_m is 10^{-4} cm/sec at a room temperature of about 20°C or slightly less.

What would q_m be if $T = -10$°C?

CHAPTER V

UNSTEADY FLOW OF HETEROGENEOUS FLUIDS

5.1 UNSTEADY FLOW WITH TWO FLUIDS

Variables pertinent to a description of unsteady flow in a porous medium containing two fluids include pressure, flux rate and permeability of each fluid along with the saturation, space coordinates and time. Fluid properties including density and viscosity are treated as constants, although for a more general analysis these properties are variable. Medium properties including porosity and maximum permeability also are treated as constants. The effective permeabilities k_w and k_{nw} are functions of p_c but are considered invariant in respect to direction; i.e., the medium is assumed to be isotropic.

5.1.1 Two-phase unsteadv flow equation - An equation relating the variables is developed by combining the two-phase flux equation, i.e., Equation 4.2, with a material balance equation. Equation 4.2 is first solved explicitly for q_{wi} to give

$$q_{wi} = \frac{k_w}{\mu_w}\left(\frac{\partial p_c}{\partial x_i} + \Delta\rho g_i\right) + q_{nwi}\left(\frac{\mu_{nw}}{\mu_w}\frac{k_w}{k_{nw}}\right) \quad . \qquad 5.1$$

A material balance equation is obtained by considering a reference volume of the bulk medium, including the solid matrix as well as a representative portion of each of the two fluids. Such a volume is of a size sufficient for the specification of porosity as explained in Section 1.3.1. The resulting material balance equation is

$$\frac{\partial q_{wi}}{\partial x_i} = - \phi \frac{\partial S}{\partial t} \quad . \qquad 5.2$$

Combining Equations 5.1 and 5.2 gives

$$- \phi \frac{\partial S}{\partial t} = \frac{\partial}{\partial x_i}\left[\frac{k_w}{\mu_w}\left(\frac{\partial p_c}{\partial x_i} + \Delta\rho g_i\right)\right] + \frac{\partial}{\partial x_i}\left[q_{nwi}\left(\frac{\mu_{nw}}{\mu_w}\frac{k_w}{k_{nw}}\right)\right] \quad . \qquad 5.3$$

Equation 5.3 is the two-phase unsteady flow equation. Problems involving displacements of one fluid by another are analyzed by obtaining solutions for this equation, with appropriate sets of initial and boundary conditions. Examples include imbibition of water in soils, infiltration, drainage of soils and the displacement of one petroleum fluid by another, e.g., oil by brine or oil by gas.

Many practical problems can be analyzed by solving simpler forms of Equation 5.3, because a number of the factors accounted for in the complete equation are not significant in all cases. A simplification often made by soil scientists for application to water-air displacements in soils is to neglect the second term on the right of Equation 5.3. This is justified by the assumption that the ratio $\mu_{nw} k_w/\mu_w k_{nw}$ is very small in most parts of the system under consideration. Since μ_w is about three orders of magnitude greater than μ_{nw}, this assumption is often justified. However, there are regions in many air-water systems where k_w is much greater than k_{nw}, in which case the assumption is not justified.

A contrasting simplification is often made by petroleum engineers when dealing with displacements of petroleum fluids, especially oil by brine. In the latter case, the second term on the right of Equation 5.3 is often much larger than the first term. The latter situation arises because the two fluids flow side by side over large distances with very small gradients of capillary pressure and saturation.

The driving force causing flow is large, and nearly the same, for both fluids. The two fluids often have comparable viscosities at the temperature existing in the aquifer. Furthermore, the slopes and the density difference are often small so that the buoyancy, $\Delta\rho g_i$, may be negligible. Under the latter conditions, the first term on the right of Equation 5.3 can be neglected with negligible error. In most oil-brine aquifers, however, there are regions where the saturation changes substantially over relatively short distances. An analysis of saturation changes in such regions requires the retention of both terms on the right of Equation 5.3.

A statement frequently made in regard to dropping the first term on the right of Equation 5.3 is that capillary, as well as buoyancy effects are disregarded. In fact, capillary effects are not entirely ignored since the ratio k_w/k_{nw} is a function of p_c. It is only the contribution of $\partial p_c/\partial x_i$ to the driving force causing flow, in addition to buoyancy, which is ignored. The effect of capillarity on k_w and k_{nw} remains implicit in the equation.

There are other simplifications which are justified for special cases. Some of these are described in the following sections.

5.2 FLOW OF WATER AND AIR IN SOILS

In the case of water-air in soils, it is customary to assume that the density and viscosity of air are negligible, and that wherever an interconnected air phase exists, the pressure of air is everywhere atmospheric. According to these assumptions, $\Delta\rho = \rho_w$ and $p_c = -p_w$. Consequently, Equation 5.3 reduces to

$$\phi \frac{\partial S}{\partial t} = -\frac{\partial}{\partial x_i}\left[\frac{k_w}{\mu_w}\left(\frac{\partial p_c}{\partial x_i} + \rho_w g_i\right)\right]$$

which is usually written as

$$\frac{\partial \theta}{\partial t} = -\frac{\partial}{\partial x_i}\left[K\left(\frac{\partial \psi}{\partial x_i} - \frac{\partial h}{\partial x_i}\right)\right] \quad , \qquad 5.4$$

where ψ is $-p_w/\rho_w g$ and h is the elevation above a datum. Equation 5.4 is known as the Richard's equation after L. A. Richards (1931). The Richards equation is further simplified where possible.

5.2.1 Horizontal linear imbibition without air resistance

- The case of horizontal linear imbibition of water into a homogeneous soil is of considerable academic interest and has been studied often. The additional simplifications made in this include neglecting the gravity force and assuming that θ is a single-valued function of ψ only. The latter assumption eliminates the possibility of analyzing flow into soils which are nonhomogeneous or soils where hysteresis may occur.

The particular case most often analyzed is that of imbibition of water into a soil with uniform water content θ_i from a source maintaining a constant water content θ at the plane where $x = 0$. It is required that θ_o satisfy the condition

$$\theta_m > \theta_o > \theta_i \quad ,$$

where θ_m is the maximum water content for the soil under consideration.

If the conditions described above are satisfied, Equation 5.4 can be simplified as

$$\frac{\partial \theta}{\partial t} = \frac{\partial}{\partial x}\left(D \frac{\partial \theta}{\partial x}\right) \quad , \qquad 5.5$$

where

$$D \equiv -K \frac{d\psi}{d\theta} ,$$

which is called the diffusivity coefficient. Defining D in terms of the total derivative of ψ is based on the assumption that ψ is a single-valued function of θ only.

Equation 5.5 is nonlinear, because D is a function of θ. A solution can be found by first reducing the equation to an ordinary form with a suitable transformation of variables, i.e., the Boltzman transformation given by

$$z(\theta) = xt^{-1/2} .$$

With this transformation, Equation 5.5 becomes

$$-\frac{z}{2}\frac{d\theta}{dz} = \frac{d}{dz}\left(D \frac{d\theta}{dz}\right) . \qquad 5.6$$

The initial and boundary conditions to be considered are:

$$\theta_o(0,t) = \theta_o ,$$

$$\theta(\infty,t) = \theta_i ,$$

$$\theta(x,0) = \theta_i .$$

Under the transformation, these become

$$\theta(z = 0) = \theta_o ,$$

$$\theta(z \to \infty) = \theta_i .$$

Since the boundary and initial conditions (as well as the starting equation) are reduced by the transformation,

$$x(\theta) = z(\theta)t^{1/2} \qquad 5.7$$

is a valid solution of Equation 5.5 for the conditions specified. This implies that the distance to a plane where θ has a specified value is proportional to the square root of time, provided that the conditions are satisfied. In this case $z(\theta)$ is a single-valued function of θ which satisfies Equation 5.6. Equation 5.7 is known as the square root of time law.

Equation 5.7 also implies that the wetted region has a saturation profile which "stretches" with time but remains similar in shape. Consequently, it seems reasonable to suppose that the cumulative inflow should also be proportional to $t^{1/2}$. In fact, this can be shown to be true based on theory which has been verified, at least roughly, by many experimenters. Figure 5.1 shows some typical results obtained by King (1964) for conditions as near as possible to those assumed for the derivation of Equation 5.7.

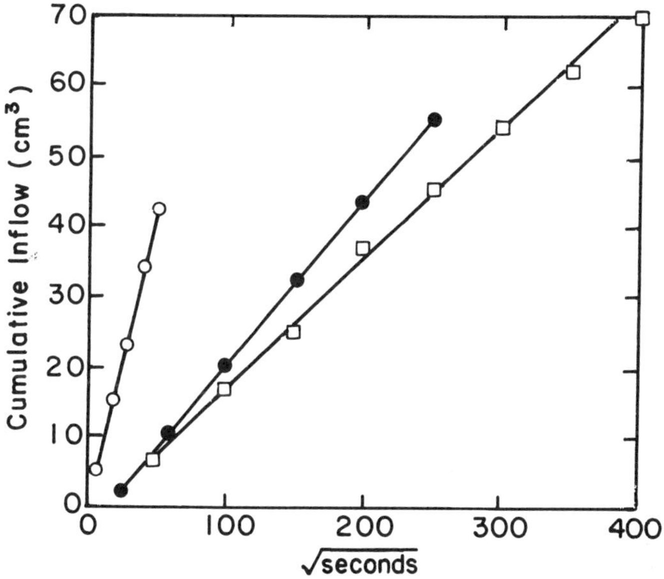

Figure 5.1. Cumulative volume of inflow as a function of $t^{1/2}$ during horizontal linear imbibition [King (1964)].

Because of the very restrictive conditions for which Equation 5.7 is derived, experimental verification is difficult. According to King (1964), the difficulties include:

(1) The soil must be packed in a homogeneous manner.
(2) The vertical dimension of the tube must be sufficiently small that the effect of gravity on the distribution of water is negligible. King stated that the vertical dimensions must be small compared to $p_d/\rho g$. In other words, the tube for a coarse-textured soil must be smaller than for a fine-textured soil.
(3) The value of ψ at the inflow boundary must be constant and larger than $p_e/\rho_w g$.

The latter condition is particularly difficult to satisfy rigorously because water must be admitted through a capillary barrier. The resistance to flow through such a barrier and through a contact region must be accounted for in some way. However, the resistance of the barrier and the contact region vary with the flow rate, and also with time because of plugging. Failure to compensate for this resistance may result in curves, such as those shown in Figure 5-1, having a small intercept on the time axis. In fact, it is practically impossible to satisfy the theoretical conditions rigorously, because Equation 5.7 implies an infinite flux rate at $t = 0$. This cannot be achieved physically, especially through a capillary barrier.

An alternative method of obtaining Equation 5.7, so that the Boltzman relationship is derived directly rather than as a trial solution, has been suggested by King and Corey. Their derivation has been presented by Bear (1972).

In the derivation of Equation 5.7 described above it was assumed, among other things, that θ is a single-valued function of ψ and that the resistance to the flow of air is negligible. In fact, an equation of the same form as Equation 5.7 can be derived without either of the latter assumptions.

To demonstrate that θ need not be a single-valued function of ψ to obtain Equation 5.7, consider the imbibition of water into a horizontal capillary tube as shown in Figure 5-1.

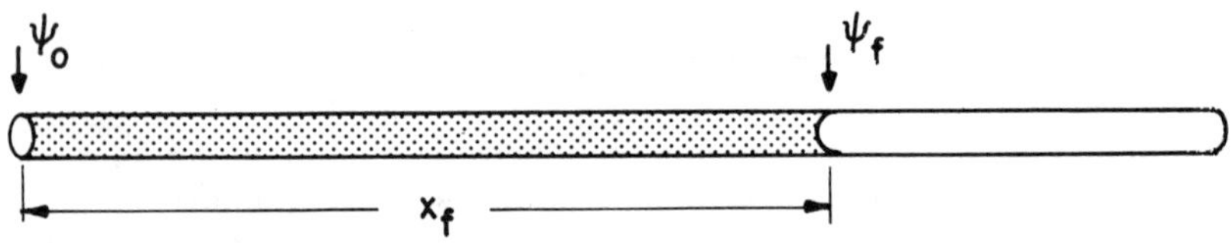

Figure 5-2. Imbibition into a horizontal capillary tube.

For the capillary tube in Figure 5-2, the flux rate q is given by

$$q = - K'\left(\frac{\psi_f - \psi_o}{x_f}\right) = \frac{c}{x_f} ,$$

where c is a constant; the subscript f refers to the wetting front, and

$$K' = \frac{r^2 \rho_w g}{8 \mu_w} ,$$

r being the radius of the tube. A continuity equation referring to the entire wetted volume of the tube is given by

$$q = c_2 \frac{dx_f}{dt} .$$

Consequently,

$$\frac{dx_f}{dt} = \frac{C}{x_f} \quad .$$

Integration results in

$$x_f^2 = C't \quad ,$$

or

$$x_f = C''t^{1/2} \quad . \qquad 5.8$$

Equation 5.8 implies that both the distance to the front and the volume of water imbibed increase in proportion to $t^{1/2}$. Note that, in this case, the value of θ behind the wetting front is constant although ψ varies from ψ_o to ψ_f. Therefore, θ is not a single-valued function of ψ. An analogous result can be obtained assuming a homogeneous tube of soil having a uniform pore size, i.e., $\lambda \rightarrow \infty$. It would seem, therefore, that the validity of the square root of time law does not necessarily imply that θ is a single-valued function of ψ.

In order for Equation 5.5 to be of any use beyond the derivation of Equation 5.7, it must be possible to measure the function $D(\theta)$. R. R. Bruce and A. Klute (1956) have presented a procedure for making such a measurement. The measurement is made by performing experiments which satisfy the conditions for which Equations 5.6 and 5.7 are valid.

Equation 5.6 is integrated between limits with respect to $z(\theta)$. The result is

$$- \int_{\theta_i}^{\theta_x} \frac{z}{2}\, d\theta = D(\theta_x)\left(\frac{d\theta}{dz}\right)_{\theta_x} - D(\theta_i)\left(\frac{d\theta}{dz}\right)_{\theta_i} \quad . \qquad 5.9$$

The last term on the right of Equation 5.9 is zero at constant time because the derivative $(d\theta/dz)_{\theta_i}$ is zero at values of x large enough to be ahead of the wetting front.

It follows that

$$D(\theta_x) = -\frac{1}{2}\left(\frac{dz}{d\theta}\right)_{\theta_x} \int_{\theta_i}^{\theta_x} z\,d\theta \quad .$$

Replacing z with its equivalent in terms of x and t, and considering a constant t results in

$$D(\theta_x) = -\frac{1}{2t}\left(\frac{dx}{d\theta}\right)_{\theta_x} \int_{\theta_i}^{\theta_x} x d\theta \quad . \qquad 5.10$$

Equation 5.10 provides the basis for the following procedure for evaluating $D(\theta)$:

(1) Obtain $\theta(x)$ at a constant t.
(2) Evaluate the integral in Equation 5.10 graphically or numerically.
(3) Calculate D at values of θ_x thereby obtaining $D(\theta)$.

The reader is referred to the paper by Bruce and Klute (1956) for details of the experimental procedure.

5.2.2 Horizontal linear imbibition with air resistance

McWhorter (1971) has shown that an equation of the same form as Equation 5.8 is obtained for linear horizontal imbibition even when resistance to displacement of the nonwetting phase is accounted for. To show this, the first term on the right of Equation 5.3 (for the 1-dimensional horizontal case) is written as for Equation 5.5. The second term is rewritten as

$$-\left[\alpha \frac{\partial q}{\partial x} + q_{nw} \frac{\partial \alpha}{\partial x}\right] \quad ,$$

where $\alpha = \mu_{nw} k_w / \mu_w k_{nw}$. Recalling that

$$\frac{\partial q_{nw}}{\partial x} = \frac{\partial \theta}{\partial t} \quad ,$$

and expanding $\partial\alpha/\partial x$, results in

$$\frac{\partial \theta}{\partial t} = f_{nw}\left[\frac{\partial}{\partial x}\left(D \frac{\partial \theta}{\partial x}\right) - q_{nw} \frac{d\alpha}{d\theta} \frac{\partial \theta}{\partial x}\right] \quad , \qquad 5.11$$

where $f_{nw} = (1 + \alpha)^{-1}$. The parameters α, f_{nw} and D are each assumed to be single-valued functions of θ.

Equation 5.11 contains two unknown dependent variables, i.e., θ and q_{nw}. Consequently, a second material balance equation is introduced involving the entire wetted portion of the soil tube as a reference volume. The cumulative volume of the nonwetting fluid is designated as Q_{nw} and is given by

$$Q_{nw} = \int_{\theta_{nwo}}^{\theta_{nwi}} x \, d\theta_{nw} \quad ,$$

where θ_{nwi} and θ_{nwo} are the initial content of the nonwetting fluid and its content at $x = 0$, respectively.

Noting that $\theta_w + \theta_{nw} = \phi$ and $d\theta_{nw} = -d\theta_w$ permits a change of variables, and differentiating with respect to t gives

$$q_{nw} = \frac{d}{dt} \int_{\phi-\theta_i}^{\phi-\theta_o} x \, d\theta \quad . \qquad 5.12$$

Equations 5.11 and 5.12 are combined to give

$$\frac{\partial\theta}{\partial t} = f_{nw}\left[\frac{\partial}{\partial x} \left(D \frac{\partial\theta}{\partial x}\right) - \left(\frac{d}{dt} \int_{\phi-\theta_i}^{\phi-\theta_o} x d\theta\right) \frac{d\alpha}{d\theta} \frac{\partial\theta}{\partial x}\right] \qquad 5.13$$

Equation 5.13 is a nonlinear partial differential equation because f_{nw}, α and D depend upon θ. The conditions considered are

$$\theta(0,t) = \theta_o \quad ,$$

$$\theta(\infty,0) = \theta_i \quad ,$$

$$\theta(x,0) = \theta_i \quad .$$

The Boltzman transformation,

$$z(\theta) = x(\theta)t^{-1/2} \quad ,$$

reduces the partial differential equation to the ordinary form

$$\frac{d}{dz} \left(D \frac{d\theta}{dz}\right) = - \frac{1}{2} \left[\bar{z} \frac{d\alpha}{d\theta} + \frac{z}{f_{nw}}\right] \frac{d\theta}{dz} \quad , \qquad 5.14$$

where

$$\bar{z} = \int_{\phi-\theta_i}^{\phi-\theta_o} z \, d\theta \quad .$$

Under the transformation, the conditions are reduced to

$$\theta(z = 0) = \theta_o$$

$$\theta(z = \infty) = \theta_i \quad ,$$

so that

$$x(\theta) = z(\theta)t^{1/2} \qquad 5.15$$

is a solution of Equation 5.13 for the conditions specified. Although Equation 5.15 has the same form as Equation 5.7 and the symbols used are the same, the two equations are not identical. The functions $x(\theta)$ and $z(\theta)$ should have different values for particular values of θ in the case where resistance to flow of the nonwetting fluid occurs, as compared to the case where resistance to the flow of the wetting fluid only is considered.

However, the development of Equation 5.15 indicates that an experimental verification of the square root of time law does not necessarily imply the validity of the Richards equation in a particular case. Evidently, the validity of the square root of time law does not depend on the resistance to air flow being negligible just as it does not depend necessarily upon the value of θ being a single-valued function of ψ.

5.2.3 Infiltration from a constant head source - The term "infiltration" designates vertical flow of water into a soil from a source at the surface, replacing air. Resistance to air flow in response to the advancing water may or may not be important. Infiltration is a common physical phenomenon and is a major process of the hydrologic cycle. Consequently, the analysis of infiltration is of more than academic interest.

Assuming that resistance to air displacement can be neglected, the downward flow can be described by Equation 5.4. For this case, Equation 5.4 becomes

$$\frac{\partial\theta}{\partial t} = - \frac{\partial}{\partial z} \left[K\left(\frac{\partial\psi}{\partial z} + 1\right)\right] \qquad 5.16$$

where z is the vertical coordinate measured positive downward, i.e., positive in the direction of flow.

The term in parenthesis in Equation 5.16 represents the combined driving force (per unit weight of water) due to the pressure gradient and gravity. The vertical component of the pressure gradient $\partial\psi/\partial z$ is a result of capillary effects due to the variation of θ with z. The gravity force per unit weight of water in a vertically downward direction is 1.0. As an aid in visualizing the way these two forces are interrelated as water infiltrates downward through the soil, an idealized model consisting of a vertical capillary tube is considered first. The idealized model is shown in Figure 5-3.

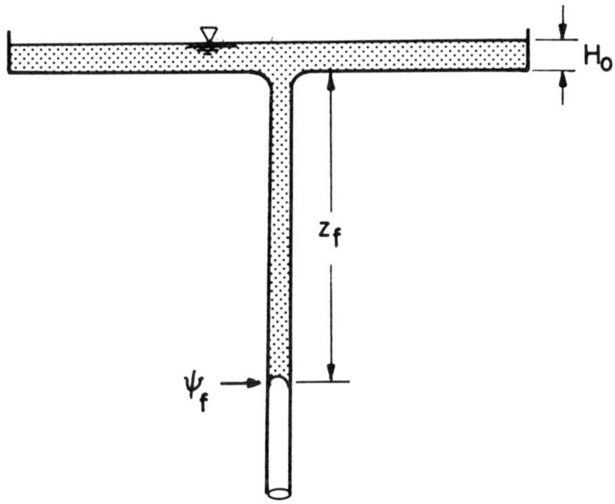

Figure 5-3. Downward imbibition into a capillary tube from a constant head source.

For the capillary tube shown in Figure 5-3, the flux rate I is given by

$$I = K'\left[\frac{H_o + z_f + \Psi_f}{z_f}\right] \qquad 5.17$$

in which K' is a constant depending on the fluid properties and the radius of the tube. H_o is the depth of ponded water which is maintained constant, z_f is the distance to the air-water interface from the source and ψ_f is the value of $-p_w/\rho g$ at the interface.

For simplicity, a case is considered for which H_o is small enough to be neglected so that

$$I = K'\left[\frac{\Psi_f + z_f}{z_f}\right] = K'\left[\frac{\Psi_f}{z_f} + 1\right] \quad ,$$

where ψ_f/z_f represents the capillary force per unit weight of water, and 1.0 is the gravity force per unit weight of water. Note that when z_f is small, the capillary force is relatively large and may be much more important than gravity. However, the capillary force decreases with z_f and eventually becomes negligible, whereas the gravity force remains constant with z_f and time. Consequently, at large times, the flux rate should approach the constant K'.

An equation in the same form as Equation 5.17, for infiltration into soils from a ponded water source, was derived by Green and Ampt (1911). For this case, K' is considered to be K_m, the maximum conductivity existing after the soil is flooded; and ψ_f has been given various interpretations. Whisler and Bouwer (1970) recommended the use of $p_e/\rho_w g$ for ψ_f, and suggested a way of measuring p_e directly in the field. This is an obvious choice since z_f should correspond to an elevation where p_c is p_e.

Equation 5.17 provides a relationship between I and z_f but does not provide any information about the time at which a particular I will occur or about the distribution of θ with respect to z and t. To obtain this information, it is necessary to introduce a continuity equation. Green and Ampt wrote such an equation by assuming that all of the water infiltrated is held by the soil above the elevation where $\psi = \psi_f$, i.e., where $p_c = p_e$. This, obviously, is a simplification of the real situation since some of the water will advance ahead of this front. Consequently, the Green and Ampt equation, in its original form, can give no information about the distribution of θ in the partially saturated region.

The continuity equation used by Green and Ampt is

$$I = (\theta_m - \theta_i) \frac{dz_f}{dt} \quad , \qquad 5.18$$

where θ_m and θ_i are the final and initial water contents respectively. Combining this with Equation 5.15 and integrating with respect to t gives

$$t \cong \frac{V}{K_m} - \frac{H_t \, \Delta\theta}{K_m} \ln\left(1 + \frac{V}{H_t \, \Delta\theta}\right) \qquad 5.19$$

in which H_t is the sum $H_o + \psi_f$, and V is the volume/area of water infiltrated in time t, i.e., $z_f \, \Delta\theta$.

The use of Equation 5.17 requires a knowledge of:

(1) The initial water content θ_i,
(2) Maximum water content θ_m,
(3) Hydraulic conductivity at θ_m, i.e., K_m,
(4) The sum $H_o + \psi_f$.

Whisler and Bouwer (1970) have suggested methods of measuring the required quantities in the field. Whisler and Bouwer (1970) and later Mein and Larson (1973) have presented results indicating that the Green and Ampt equation agrees well with measured data. Considering the assumptions made in its derivation, this equation would be expected to have greater validity for soils with a relatively narrow range of pore size. In such soils, the quantity of water advancing beyond the front where $\psi = \psi_f$ should be minimal, especially at early times, and the resistance to air flow ahead of the front also should be minimal.

Philip (1957) was among the first to solve Equation 5.16 for the case of infiltration from a constant head source. He first wrote this equation in the θ-dependent form

$$\frac{\partial \theta}{\partial t} = \frac{\partial}{\partial z} \left(D \frac{\partial \theta}{\partial z}\right) - \frac{\partial K}{\partial z} , \qquad 5.20$$

where z is the vertical coordinate, positive downward. For the case considered, the conditions are:

$$t = 0,\ z > 0,\ \theta = \theta_i$$

$$t \geqq 0,\ z = 0,\ \theta = \theta_o$$

in which θ_i is the initial water content and θ_o is a water content maintained at the surface by a constant ψ at $z = 0$.

Equation 5.20 is converted into a z-dependent form by multiplying each term by $\partial z/\partial \theta$ and using the identity

$$(\partial \theta/\partial t)_z \cdot (\partial z/\partial \theta)_t = -(\partial z/\partial t)_\theta$$

so that

$$-\frac{\partial z}{\partial t} = \frac{\partial}{\partial \theta} \left(D \frac{\partial \theta}{\partial z}\right) - \frac{\partial K}{\partial \theta} , \qquad 5.21$$

which is subject to the conditions pertaining to Equation 5.20.

The solution of Equation 5.21 is considered to be a perturbation of the solution of

$$- \frac{\partial z'}{\partial t} = \frac{\partial}{\partial \theta} (D \frac{\partial \theta}{\partial z'}) \quad , \qquad 5.22$$

where z' is taken as a first estimate of z. The solution of Equation 5.22 is analogous to Equation 5.15, i.e.,

$$z' = \alpha(\theta)t^{1/2} \quad . \qquad 5.23$$

Subtracting Equation 5.21 from Equation 5.22 gives

$$\frac{\partial y}{\partial t} = \frac{\partial}{\partial \theta} (D \frac{\partial \theta}{\partial z'} \frac{\partial y}{\partial z}) + \frac{\partial K}{\partial \theta} \quad , \qquad 5.24$$

where

$$y = z - z' \quad .$$

The approximation

$$\frac{\partial y}{\partial z} \approx \frac{\partial y}{\partial z'}$$

leads to

$$\frac{\partial y'}{\partial t} \approx \frac{\partial}{\partial \theta} [D(\frac{\partial \theta}{\partial z'})^2 \frac{\partial y'}{\partial \theta}] + \frac{\partial K}{\partial \theta} \quad , \qquad 5.25$$

where y' is used in place of y because of the approximation.

The transformation

$$\chi = y't^{-1}$$

leads to

$$\chi = \frac{d}{d\theta} (P \frac{d\chi}{d\theta}) + \frac{dK}{d\theta} \qquad 5.26$$

in which

$$P(\theta) = D(\frac{d\theta}{d\rho})^2 \quad .$$

From Equation 5.23

$$\alpha(\theta) = z't^{-1/2} \quad . \qquad 5.27$$

Integrating Equation 5.26 with respect to θ gives

$$\int_{\theta_i}^{\theta} \chi \, d\theta = P \frac{d\chi}{d\theta} + (K - K_i) \quad , \qquad 5.28$$

where K_i corresponds to θ_i. The conditions are

$$\theta = \theta_i, \quad \chi = 0 \quad .$$

Equation 5.28 is solved numerically for this condition. A new residual error

$$y'' = y - y'$$

is introduced and the procedure is repeated with new residuals as often as required to give the necessary accuracy.

The solution finally is developed as a power series in $t^{1/2}$:

$$z(\theta) = \alpha t^{1/2} + \chi t + \psi t^{-3/2} + \ldots + f_m(\theta)t^{m/2} + \ldots \qquad 5.29$$

where the coefficients of $t^{m/2}$ are functions of θ which are solutions of a series of ordinary equations that can be solved numerically. For details of the method of calculation, the reader is referred to the paper by Philip (1957).

According to Philip (1957), within the range of variables most often encountered, only the first few terms of Equation 5.29 are needed since the series converges rapidly. For practical applications, the first two terms on the right are often considered sufficient. With this simplification, Equation 5.29 can be shown to be in approximate agreement with the Green and Ampt equation where $z(\theta) = z(\theta_m)$. In this case,

$$I \simeq \Delta\theta \frac{dz(\theta_m)}{dt} \quad .$$

At large times, $I \rightarrow \chi$ where χ is an approximation of K_m. At very early times, the first term is more important so that the infiltration tends to proceed as it would for the horizontal case, indicating that capillary forces are predominant at early times.

5.2.4 Infiltration with air resistance - The analysis of infiltration from a constant head source presented in the preceding section assumes that the resistance to flow of air ahead of the invading water is negligible so that the pressure of air is everywhere atmospheric. It is not clear that this is always a valid approximation.

Theoretically, an analysis could be made without this approximation by solving Equation 5.3 for the 1-dimensional vertical case. However, it usually is expedient to make further simplifications. For example, Morel-Seytoux (1973) and

Morel-Seytoux and Khanji (1974) have presented an analysis employing the following procedure:

An explicit relationship for a total flux q_t is obtained by adding the expressions for the fluxes q_w (water) and q_a (air). The result is

$$q_t = -\frac{k_w}{\mu_w}\frac{\partial p_w}{\partial z} - \frac{k_a}{u_a}\frac{\partial p_a}{\partial z} + \frac{k_w}{\mu_w}\rho_w g + \frac{k_a}{\mu_a}\rho_a g \quad , \qquad 5.30$$

in which q_t and z are taken as positive downward.

Noting that $p_a = p_w + p_c$ and neglecting $\rho_a g$ gives, after rearrangement,

$$\frac{q_t}{\Lambda} = -\frac{\partial p_w}{\partial z} - (1 - f_w)\frac{\partial p_c}{\partial z} + \Delta\rho g f_w \quad , \qquad 5.31$$

where Λ is $(k_w/\mu_w + k_a/\mu_a)$ and f_w is $[1 + (\mu_w k_a/\mu_a k_w)]^{-1}$.

Equation 5.31 is integrated with respect to z keeping q_t constant, since q_t does not vary independently with z. The result is

$$q_t = \frac{(p_w)_1 + (p_c)_1 - (p_a)_2 + \int_{p_{c_1}}^{p_{c_2}} f_w dp_c + \Delta\rho g \int_{z_1}^{z_2} f_w \, dz}{\int_{z_1}^{z_2} \frac{dz}{\Lambda}} \quad , \qquad 5.32$$

in which the subscripts 1 and 2 refer to the positions over which p_w and p_c are integrated. Equation 5.32 theoretically could be applied to either infiltration or drainage since nothing is assumed about which phase is displaced.

For most cases, a solution of Equation 5.32 requires numerical calculations but for the case of infiltration from a constant head source, the expression for q_t reduces to a simple form. When water is ponded to a depth H at the soil surface, the soil at the surface becomes fully saturated, except for entrapped air. Also, the value of p_c at the soil surface is zero. The fully saturated region extends to a depth z_f, meaning the depth to the wetting front below which $S < S_m$ and above which $S = S_m$. Within the fully saturated region, only water flows so that f_w equals 1.0 and q_t is q_w. The value of q_w is called the infiltration rate and is designated by I.

In many cases, it is reasonable to assume that the air pressure ahead of the wetting front is atmospheric at all times. This is not a valid assumption, however, if a shallow water table or some other restriction to the free escape of air exists below the wetting front. Above the wetting front, k_a is zero so that Λ is equal to k_m/μ_w where k_m is the value of k_w at S_m. With these simplifications, Equation 5.32 becomes

$$I = K_m[H + \int_o^{h_f} f_w \, dh_c + \int_o^{z_f} f_w \, dz] \, z_f^{-1} \quad , \qquad 5.33$$

in which all terms have been converted to "heads" by dividing Equation 5.32 by $\rho_w g$, and K_m is $k_m \rho_w g/\mu_w$. Since $f_w = 1.0$ above the wetting front, this reduces to

$$I = K_m(H + h_f + z_f)z_f^{-1} \qquad 5.34$$

in which h_f is the value of $p_c/\rho_w g$ at z_f. Note that except for the symbols used, Equation 5.34 is identical to the Green and Ampt Equation 5.17.

Where air resistance cannot be ignored, the integration of Equation 5.32 is more complex. Furthermore, if one needs to determine I or S as a function of t, it is necessary to introduce a continuity equation. The continuity equation introduced by Morel-Seytoux and Khanji (1974) is

$$\frac{\partial q_w}{\partial z} = - \frac{\partial \theta}{\partial t} \quad . \qquad 5.35$$

Equation 5.35 is integrated with respect to z over the entire soil depth in which a significant amount of infiltrated water exists. To facilitate the integration, the authors assumed a saturation profile given by the Welge equation (see Section 5.3.2) which neglects gravity and gradients of p_c. Their rationale for making this simplification is that the quantity of most concern, i.e., I is very insensitive to the exact water content profile selected.

Based on these assumptions, Morel-Seytoux and Khanji (1974) obtained

$$t = \frac{\beta}{K_m} \{V - (H + H_c)\, \Delta\theta\, \ell n[1 + \frac{V}{(H + H_c)\Delta\theta}]\} \quad , \qquad 5.36$$

where V is the volume of water per unit area infiltrated in time t;

$$\beta = \frac{\Delta\theta}{\mu_w} \int_{\theta_m}^{\theta^-} \frac{f''_w\, d}{(k_{rw}/\mu_w) + (k_{ra}/\mu_a)} \quad ;$$

θ^- is the water content just upstream of the "discontinuity" front; f''_w is $d^2f/d\theta^2$, and

$$H_c = \int_o^{h_c^-} f_w\, dh_c \quad ,$$

in which h_c^- is the value of h_c corresponding to θ^-. The term β/K_m results from the evaluation of the integral in the denominator of Equation 5.32, retaining the term k_a/μ_a in the expression for Λ.

Equation 5.36 should provide a better estimation of $V(t)$ than Equation 5.19 where the resistance to air flow is not negligible, and particularly where the transition region of the saturation profile is relatively long. Besides the factor of air resistance, there is also the factor of the water stored in the transition region which is accounted for in an approximate manner by Equation 5.36. Note that if air resistance and water stored in the transition region are neglected, Equation 5.36 is identical with Equation 5.19.

5.2.5 Infiltration with air compression - Another case of infiltration from a ponded surface source occurs when the soil contains an underlying water table or some other restrictive layer which delays the escape of air and results in compression of the air ahead of the wetted front. The effect of this is an increase in the air pressure resulting in a decrease in infiltration rate, frequently followed by an abrupt breakthrough of the air to the surface. Youngs and Peck (1964), Peck (1965) and McWhorter (1971) are among those who have studied this situation experimentally and theoretically.

The observations of these investigators indicate that the threshold pressure at which air breaks through to the surface is significantly larger than that which is indicated from the primary loop of a wetting $p_c(\theta)$ curve. Evidently, a region of $\theta \simeq \theta_m$ moves into the soil during compression, and at the time of breakthrough, there is a slight desaturation of this region so that the threshold pressure observed is one that occurs on a desaturation loop starting with θ_m.

Prior to breakthrough, there is a substantial but gradual reduction in infiltration rate. After breakthrough, a counterflow of air begins, accompanied by an immediate increase in the infiltration rate. At this time there is often a sharp decrease in air pressure which is thought to be caused by a disturbance of the soil at the time of air breakthrough. McWhorter (1971) tested this hypothesis by conducting an independent experiment with a column of consolidated Berea sandstone in which no decrease in air pressure was observed at the time of breakthrough.

Experimental results obtained by McWhorter (1971) are shown in Figures 5-4 and 5-5, illustrating the phenomena described above. The cumulative inflow into the columns was predicted by applying a theory somewhat analogous to that leading to Equation 5.32. However, the equation used and the method of its solution are beyond the scope of this text since the problem involves the compression of a nonwetting phase, which is not accounted for in Equation 5.3.

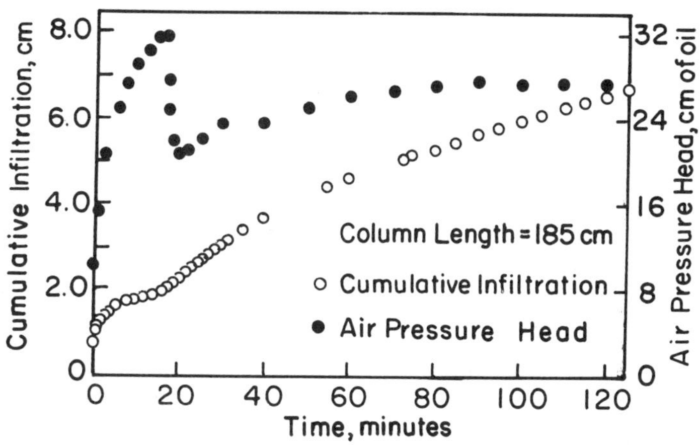

Figure 5-4. Infiltration into a column of Poudre sand (of 185 cm length) with air compression followed by air counterflow.

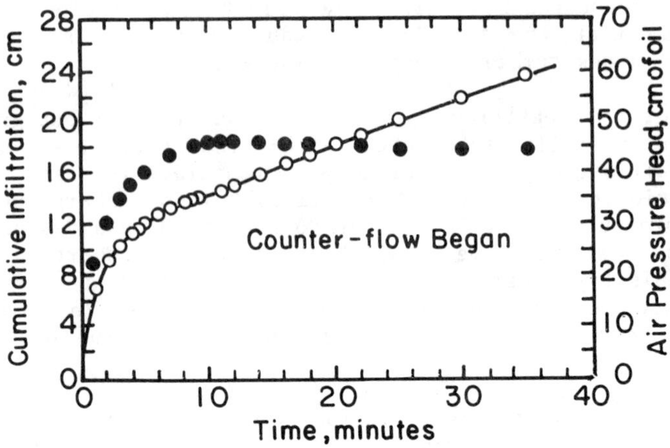

Figure 5-5. Infiltration into a column of Berea sandstone (of 17.4 cm length) with air compression followed by air counterflow.

5.2.6 Constant rate infiltration - In the preceding sections dealing with infiltration, a condition of constant water pressure or constant water content is imposed at the soil surface. Imposing a constant value of q_w at the surface is of equal interest. However, the latter condition has not received as much theoretical study as the former case. Perhaps this is because the constant rate case is not amenable to the type of solutions that can be used for the constant head case. Several cases of constant rate are of interest:

(1) Infiltration into a semi-infinite homogeneous profile with no restrictive layer, at a rate smaller than K_m so that no ponding occurs;
(2) Infiltration into a similar profile at a rate exceeding K_m so that ponding occurs after a period of time;
(3) Infiltration into a soil profile with a restriction at some depth, such as a water table, producing a counterflow of air toward the surface.

The first of these cases has been investigated by Rubin and Steinhardt (1963), using numerical methods for solving Equation 5.20 subject to the constant rate boundary condition. Parlange (1972) obtained an analytical solution of the same problem after first rewriting Equation 5.20 in the z-dependent form, i.e., Equation 5.21, as Philip (1957) had written it earlier for the constant water-content condition. Parlange obtained his solution by using a method of successive approximations described by Ames (1965), and employed by McWhorter (1971 and 1976) for solving 2-phase flow problems.

The results of Parlange's analytical solution agree with the numerical results of Rubin and Steinhardt and confirm the qualitative observations noted by numerous experimental investigators.

For a case in which q_w at the surface is less than K_m and resistance to air flow is a negligible factor, the qualitative observations are as follows:

(1) After a period of time such that water has infiltrated to a substantial depth, a region of the soil profile starting at the surface exists which is wetted to a nearly constant water content and in which the capillary pressure is nearly constant. In this region, the flow is practically steady and the driving force is mainly gravity. This is called the "transmission" region.

(2) A "transition region" exists, below the transmission region, which has a water content decreasing toward the lower end of the region. Within this region, the driving force is predominantly $\partial p_c/\partial z$.

(3) The length of the transition region is relatively longer in soils having a wider range of pore sizes, i.e., a smaller value of λ.

(4) The value of θ in the transmission region depends on the ratio of q_w/K_m.

(5) The value of K_w in the transmission region approaches q_w.

For cases in which q_w at the surface is greater than K_m, the problem is more difficult because a region of θ close to θ_m develops quickly at the surface. Furthermore, the steady inflow rate cannot be maintained. Ponding occurs where $\theta = \theta_m$ and the problem of practical significance is to determine the time at which ponding takes place. The high value of θ near the surface creates a situation in which resistance to air flow may become important.

The latter problem has been studied by Smith (1972), by Mein and Larson (1973) and by Morel-Seytoux (1975). Morel-Seytoux derived an approximate analytical equation for the time of ponding t_p. His equation is a further extension of the theory upon which Equation 5.32 is based. The expression for t_p resulting from this theory is given by

$$t_p = \frac{(\theta_m - \theta_i)H_c}{(1 - f_i)r} [e^{(\beta\hat{r}-1)^{-1}} - 1] \qquad 5.37$$

in which r is the rainfall rate (q_w at $z = 0$), $\hat{r}$ is r/K_m, f_i is f_w corresponding to θ_i, β is a factor which results from the evaluation of the integral in the denominator of Equation 5.32 and H_c is as defined for Equation 5.36, except that p_c at $z = 0$ is not assumed to be zero. For cases in which $\theta_i = \theta_r$, $f_i \to 0$.

The restricted air-flow case may result from long lateral escape routes, high water tables or soil strata that are impermeable to air because of high water content or dense structure. This case has been investigated by McWhorter (1971) and McWhorter (1976). His formulation of the problem begins with the unsteady 2-phase flow equation written for the case of 1-dimensional vertical flow. This equation is converted to the θ-dependent form. It is assumed that air is totally restricted from flowing downward or laterally, so that air can escape only upward through the wetted soil region at a rate equal to the flux rate of water into the soil surface, i.e., $q_{nw} = - q_w$. Other conditions imposed for McWhorter's (1976) solution included $q_w = q_o$ (constant), at $z = 0$ and $t \geqq 0$; and $\theta = \theta_i$ at $z > 0$ and $t = 0$. The equation was solved by a method of successive approximations similar to that used by Parlange (1972) for solving Equation 5.20.

McWhorter (1976) calculated numerical results for his equation as well as that of Parlange (1972) using a Brooks-Corey (1966) relationship among the variables k_w, k_a and p_c, and assuming an arbitrary value of λ and q_o. By comparing the profiles of $\theta = f(z)$ obtained with and without taking air resistance into account, he noted that at early times, air resistance has little effect so that the profiles are nearly the same. At longer times when θ at the surface gets closer to θ_m, the effect of air resistance becomes apparent. At this time, there is a more rapid increase in θ at the surface and ponding may occur shortly after, even when q_o is less than K_m.

Parlange's (1972) equation does not predict any ponding when q_o is less than K_m. However, when q_o is greater than K_m, Parlange's equation predicts a larger time for ponding than is the case with air resistance. On the other hand, McWhorter's (1976) equation probably underestimates the time for ponding, because it assumes that air is incompressible and that $q_o = q_w$. In a real case, especially where the water table is at a substantial depth, the ponding time may be intermediate between that predicted by the two equations. In the limit, where the depth to the water table is large, Parlange's equation should be accurate; and at small times, both equations should be valid.

The effect of air counter flow has sometimes been observed in field situations, particularly in fields under border irrigation [Dixon and Linden, 1972]. The magnitude of the effect can be expected to vary with the depth to a water table, the width of borders, the period of time for which the water source persists and on soil properties, among other factors.

The behavior described above for constant rate infiltration, where air resistance is negligible, provides a convenient method for determining K_w as a function of θ or p_c on a wetting cycle. The method depends upon the existence of a transmission region at the upper portion of a soil column when water is introduced at a constant rate at the surface. This method has been employed by Davidson et al. (1963) for obtaining $K_w(\theta)$ and by Anat et al. (1965) for obtaining $K_w(p_c)$.

The method employed uses equipment identical to that described in Section 4.2.1 for determination of K_w on the drainage cycle by either the "long" or "short"-column procedure. The only difference is that the first measurements are made before the column is fully saturated. Water is admitted to the top of the column at a small steady rate. After the wetting front has reached the bottom of the column where the outflow is withdrawn, the upper part of the column is at some constant θ and p_c which are measured. The piezometric gradient within the transmission region is approximately 1.0 and the value of K_w is some constant equal to the inflow rate q_w. The inflow rate is then increased by small increments and a set of measurements is made for each increment until θ approaches θ_m and K_w approaches K_m.

5.2.7 Linear drainage - Drainage is a process in which air replaces water from soils. A typical field situation of practical interest in this regard is an unconfined aquifer with a falling water table. The flow that occurs in the soil above the water table may be almost linear if the water table is relatively horizontal and remains that way while the water table is lowered. Another case of interest is the redistribution of water toward drier soil below (following a heavy rain or irrigation) where the water table (if one exists) is too deep to affect the flow.

Soil-water engineers, however, are primarily interested in the water table case, because it is for this case that artificial drains may be required. Frequently, artificial drains consisting of perforated tubes are placed in a parallel pattern at some depth below the water table. This creates a 2-dimensional flow pattern after a rain or irrigation. The water table slopes downward toward the drains creating a horizontal component of flow both below and above the water table.

In the usual field situation, water tables fall gradually toward an equilibrium position following a rain or irrigation. The downward component of the water flux in the soil above the water table may be controlled by the rate at which water can move laterally toward a drain, thus permitting the water table to fall. Although, a sudden incremental lowering of a water table rarely occurs in the field, it often is imposed in laboratory experiments or as a boundary condition for analytical solutions. Consequently, this type of drainage process also has been given considerable theoretical attention. Sometimes a study of drainage following sudden drawdowns leads to insights that permit an understanding of certain field behavior.

The theory used to analyze drainage differs only in details of solution from that employed to study infiltration. Various investigators have idealized the problem as one of single-phase flow of water and have employed the Richards equation in their analyses. More recently the problem has been analyzed as a displacement process involving 2-phase flow.

Vertical drainage of a soil column (which is initially fully saturated except for trapped air) is a problem of mainly academic interest, but a study of this problem may provide insights into aspects of several real problems. The drainage under consideration occurs when the plane of atmospheric pressure in the soil water is suddenly lowered from one stationary position to another. If the water table is initially at the surface, however, there may be essentially no drainage until the water table is lowered to a depth equal to $p_e/\rho_w g$ below the surface.

The simplest way possible of analyzing this problem is entirely analogous to the Green and Ampt analysis of infiltration. The method might be called an "upside-down" Green and Ampt approach. It is assumed that an abrupt drainage of the pore space takes place when the pressure of the liquid is lowered below atmospheric pressure to some critical pressure. The critical pressure represents the value of p_e for the soil under consideration. Below the level where p_e exists, the soil is fully saturated and above this point, the soil is assumed to be drained to a water content equal to θ_r. The existence of a transition region is ignored. Ligon et al. (1962) and Youngs (1960) have presented typical analyses based upon this assumption.

The analysis of Youngs (based upon the "upside-down" Green and Ampt approach) leads to

$$\frac{Q}{Q_\infty} = 1 - e^{-q_o t/Q_\infty} \qquad 5.38$$

in which Q is the cumulative outflow in time t, Q_∞ is the outflow at $t \to \infty$, and q_o is the initial flow rate. This equation represents only the first term in a series solution. However, according to Jensen and Hanks (1967), Equation 5.38 is reasonably accurate for Q/Q_∞ up to about 0.6, after which it begins to overestimate the outflow volume.

Gardner (1962) derived another approximate solution for the same boundary conditions. His equation was obtained by assuming that the average conductivity $\bar{K}$ between the surface of a column and the top of the fully saturated region remains constant during the drainage process. He also employed an adjustment (for the increasing length of partially saturated medium) which was proposed by Miller and Elrick (1958). His equation is

$$\frac{Q}{Q_\infty} = 1 - \frac{8}{\pi^2} e^{-(\alpha)^2 \bar{D} t/L} \qquad 5.39$$

in which $\bar{D} = \bar{K}L^2/Q_\infty$, L being the length of the column and $(\alpha)^2$ is obtained from the ratio of impedance of the partially saturated region to that of the fully saturated region. At the beginning of drainage, $(\alpha)^2$ has the value of $\pi/4$. Equation 5.39 like Equation 5.38, represents the first term of a rapidly converging series.

In comparing the values of Q/Q_∞ predicted by Equation 5.39 with experimental data, Gardner found he could select a value of $\bar{K}$ which would allow the equation to fit the data fairly well. Jensen and Hanks (1967) tried to find a simple way of predetermining $\bar{K}$ because, in a practical case, this would be necessary. The methods they tried for determining $\bar{K}$ did not result in satisfactory agreement between Equation 5.39 and experimental data. They suggested a better method of determining $\bar{K}$ would improve the accuracy of Equation 5.39.

Jensen and Hanks (1967) employed a numerical method developed by Hanks and Bowers (1962) to solve the Richards equation for vertical drainage in one dimension with a plane of atmospheric pressure at the base of a column. They used a computer to obtain Q/Q_∞ as a function of time and pressure as a function of depth and time. Typical results are shown in comparison with experimental data in Figures 5-6 and 5-7.

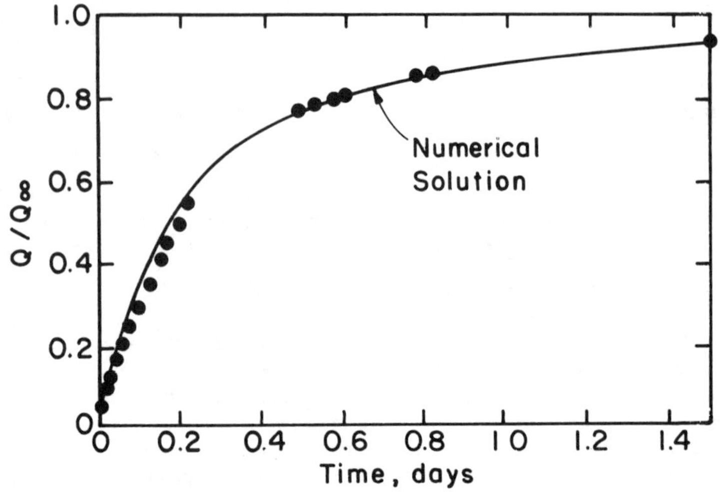

Figure 5-6. Comparison of laboratory data with numerical solution of Richards equation for vertical drainage [Jensen and Hanks (1967)].

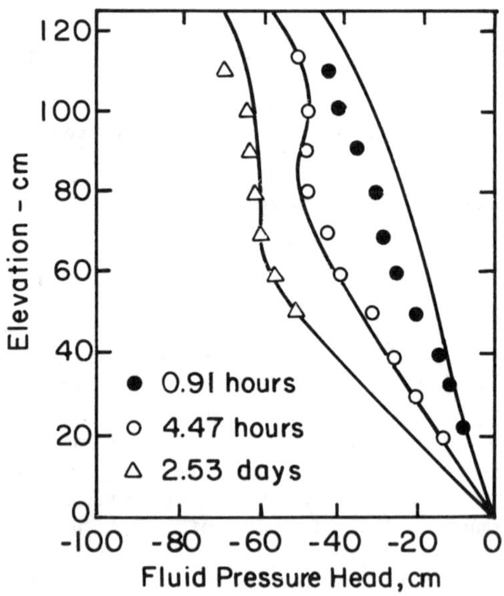

Figure 5-7. Comparison of laboratory data with numerical solution of Richards equation for vertical drainage [Jensen and Hanks (1967)].

The functional relationship among S, p_c and K_w needed for the numerical solution of the Richards equation was determined independently by steady-state techniques similar to those described in Section 3.10.3. The results of Jensen and Hanks, therefore, seem to confirm the validity of the Richards equation for an unsteady drainage situation. However, the Richards equation does not rigorously apply to that portion of the draining column which is fully saturated. This is because the function D, equal to $-Kdp_c/d\theta$ does not exist in the fully saturated region. Jensen and Hanks sidestepped this problem by assigning an arbitrarily small constant value to $dp_c/d\theta$ where $p_c < p_e$. They found that the computed drainage rates were very insensitive to the exact value of $dp_c/d\theta$ chosen, provided that the value was small.

It would appear from the numerical results of Jensen and Hanks that the 2-phase approach is not necessary in the case of drainage. However, consideration of drainage from the point of view of a 2-phase flow process may provide insights in respect to the interrelation between drainage and infiltration.

Morel-Seytoux (1975) is one of the first to have studied vertical drainage as a 2-phase flow problem. His approach is an application of the theory discussed in Section 5.2.4 which leads to Equation 5.32. Equation 5.32 applies to any 2-phase vertical displacement process, infiltration or drainage.

A physical model which can be analyzed by Equation 5.32 is an unconfined aquifer with a fluctuating water table at a depth D below a dry surface layer. It is assumed that the aquifer is homogeneous and that any flow that results from a lowering of the water table is vertically downward. It is assumed also that at a particular time, the water table has been at a fixed depth D_1 long enough for a practically static saturation profile to have developed. At time $t = 0$, the water table is lowered to a new position D_2 below the surface. It is desired to predict the rate at which water drains to the new water table.

For this situation, it is convenient to evaluate Equation 5.32 between the dry layer where $h_c \to \infty$ and any water table level D where $h_c = 0$. In this case, p_{a1} (given by the sum $p_{w1} + p_{c1}$) is equal to atmospheric pressure, which is assigned a value of zero. Also p_{a2}, the air pressure at D, is zero because there is no continuous air phase at that level. Consequently, Equation 5.32 reduces to

$$q_t = \frac{\rho_w g \int_{\infty}^{0} f_w \, dh_c + \rho_w g \int_{0}^{D} f_w \, dz}{\int_{0}^{D} \frac{dz}{\Lambda}} \quad . \qquad 5.40$$

The total flux q_t in Equation 5.40 represents both the rate of drainage and the rate of air entrance through the soil surface, the fluid having been assumed to be incompressible. Because of the relatively small deviation of pressures from atmospheric, this assumption is expected to be a valid approximation.

When the water table is at D_1, the system is assumed to be static, Actually, since the surface is air dried, a state of true static equilibrium cannot exist. However, if the immediate vicinity of the air dried layer is neglected, the soil water may be sufficiently close to the static condition to justify such an assumption. This is because the rate of water diffusion through the air-dry soil layer has a negligible effect on the water in the wet soil below. Furthermore, the value of neither of the integrals in the numerator of Equation 5.40 is affected by carrying out the integration from the water table to a position where the saturation is residual rather than to the dry surface, because the value of f_w is zero in the dry region.

Since for a static case $dh_c = dz$, Equation 5.40 satisfies the condition that q_t is zero for the water table at D_1; that is, for $t < 0$. After the water table is suddenly lowered to a depth D_2, the sum of the integrals must undergo an instantaneous increase from zero to a value given by $\rho_w g(D_2 - D_1)$. This change comes about from the increase in the second integral, which represents the gravitational force. The value of the capillary force, represented by the first integral, is assumed to undergo no instantaneous change.

At later times, as drainage proceeds, the gravitational force gradually diminishes owing to the reduction of the head of soil water above the water table. According to Morel-Seytoux (1975), the unbalanced driving head H_d at any $t > 0$ is given by

$$H_d = \rho_w g[(D_2 - D_1) - \frac{W}{\phi_e}]$$

in which W is the cumulative volume of water/area drained from the column at time t and ϕ_e is the effective porosity. Clearly, this expression for H_d is a valid approximation only for cases in which D_1 is sufficiently large that the specific yield no longer changes with D. The depth for which this condition is satisfied is discussed in Section 2.5.1.

For water table depths large enough that the specific yield S_y is constant, an approximation for the rate of drainage is given by

$$q_w \cong \frac{\rho_w g[\Delta D - W/\phi_e]}{\int_0^{D_2} \frac{dz}{\Lambda}} \quad . \tag{5.41}$$

The denominator in Equation 5.41 may be interpreted as a total resistance factor that includes resistance to flow of air as well as flow of water.

The evaluation of the integral in the denominator of Equation 5.41 requires the introduction of a continuity relationship. The relationship introduced by Morel-Seytoux (1975) is an approximation written in terms of the air content θ_a,

$$\frac{\partial \theta_a}{\partial t} \cong - q_t \frac{df_a}{d\theta_a} \frac{\partial \theta_a}{\partial z} \quad , \tag{5.42}$$

where f_a is $[1 + (\mu_a k_w/\mu_w k_a)]^{-1}$. The variable f_a is an approximation of the ratio q_a/q_t. The substitution of f_a for this ratio in Equation 5.40 is equivalent to assuming $\partial p_c/\partial z = \rho_w g$ which is strictly true only for a static distribution of pressure. The advantage of using f_a is that its derivative with respect to θ_a can be evaluated from relative permeability data only. The retention of the term $\partial p_c/\partial z - \rho_w g$ introduces great mathematical difficulties.

The assumption of Morel-Seytoux in neglecting the term $(\partial p_c/\partial x_i - \Delta\rho g \sin\beta)$ to obtain a saturation profile is not different from an assumption frequently applied in petroleum reservoir calculations. The validity of this assumption for a soil-water flow situation (in which $\partial p_c/\partial z$ may be very large) is not entirely clear. Apparently, however, the value of the resistance integral is relatively insensitive to the saturation profile assumed so that the resulting expression for q_w, (that is, dw/dt) should be reasonably accurate. The expression derived by Morel-Seytoux is

$$\frac{dw}{dt} = \frac{K[\Delta D - W/\phi_e]}{D^* - \beta(W/\phi_e)} \tag{5.43}$$

in which D^* is given by

$$D^* = \int_0^\infty \frac{dh_c}{\Lambda}$$

and β is the correction factor used in the infiltration Equation 5.36 discussed in Section 5.2.4. Integration of Equation 5.43 with respect to time yields

$$\frac{Kt}{\alpha} \cong W - (\frac{D^*}{\alpha} - \Delta D)\ \phi_e\ \ln\ (1 - \frac{W}{\phi_e\ \Delta D}) \tag{5.44}$$

provided $\Delta D < D^*/\alpha$. For larger drawdowns,

$$W \cong \Delta D\Phi_e [1 - \exp \frac{-Kt}{\phi_e (D^* - \alpha\Delta D)}] \quad . \tag{5.45}$$

In Equation 5.44 and 5.45, α is given by

$$\alpha = \phi_e\ [\frac{\bar{f}_a}{\bar{\theta}_a} - \int_{\bar{\theta}_a}^{\theta_{ai}} - \frac{1}{\Lambda} \frac{\partial^2 f_a}{\partial \theta_a^2}\ d\theta_a]$$

in which $\bar{f}_a$ and $\bar{\theta}_a$ refer to quantities evaluated at the displacement front.

The value of $\bar{\theta}_a$ is determined by a material balance calculation. A discontinuity in saturation at the displacement front is postulated, and the air volume above the front must equal the volume of water that was displaced [see Figure 5-8]. The air content at the front in this case (or the water content at the front in the case of an infiltration process) can be found by a graphical process which avoids a trial and error solution. The graphical construction, which is called the Welge construction, is shown in Figure 5-9. The justification for the assumption of a saturation distribution based on Equation 5.40 and the Welge construction is that the resistance integral is relatively insensitive to the exact profile

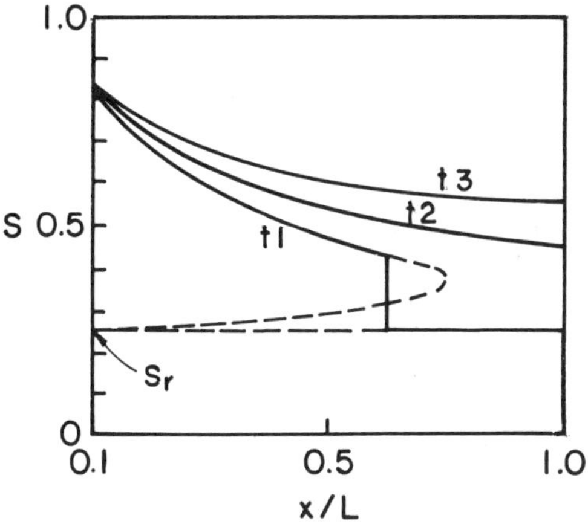

Figure 5-8. Saturation profiles for a linear displacement.

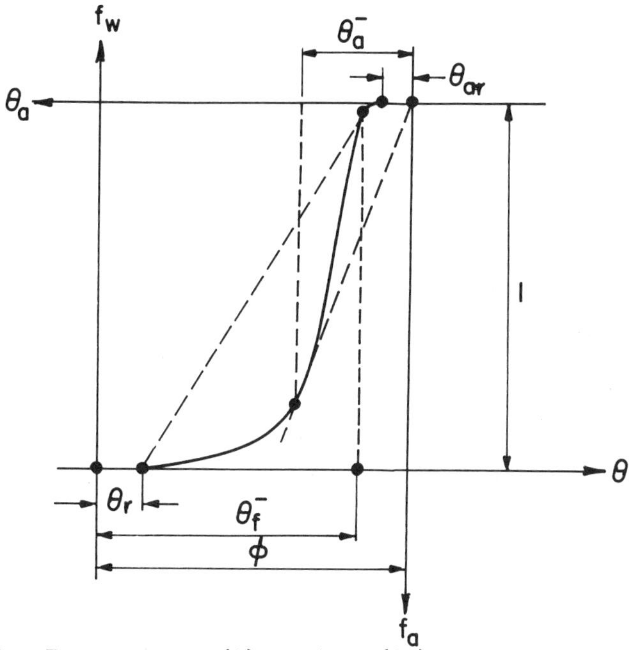

Figure 5-9. Functions $f_w(\theta)$ and $f_a(\theta_a)$ for cases of infiltration and drainage [Morel-Seytoux (1975)].

assumed. Consequently, the flow rate predicted should be accurate. However, another method should be used to obtain an accurate saturation profile.

5.2.8 Two-dimensional drainage - A problem of considerable interest to soil-water engineers involves 2-dimensional transient flow toward parallel relief drains. The physical model to be considered is analogous to that described in Section 4.4 except that a period is considered in which no recharge occurs and the water table falls. The problem is to determine the position of the water table at the midpoint between drains as a function of time [see Figure 5-10]. The objective is to determine whether or not a particular depth and spacing of drains will satisfy design criteria with respect to lowering of the water table after heavy irrigations or flooding.

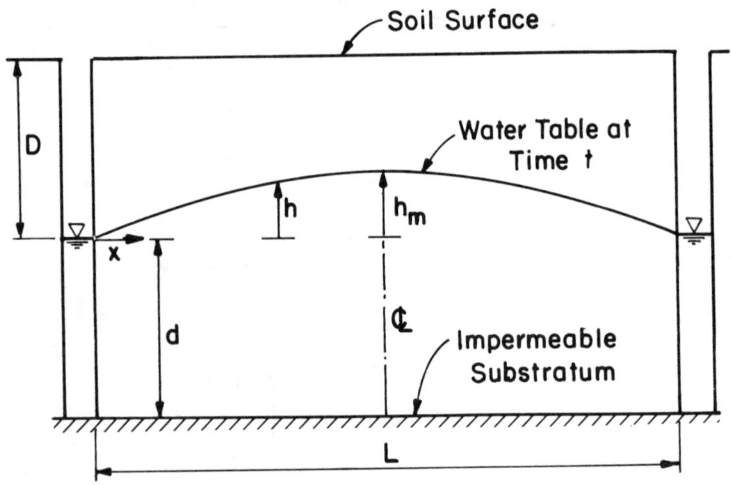

Figure 5-10. Physical model for transient flow towards parallel drains. [McWhorter and Duke (1976)].

The traditional procedure is to ignore both the flow and storage of water above the water table. This approach has led to the well-known and frequently used equation of Glover which has been described by Dumm (1964 and 1967). Use is made of the Dupuit-Forchheimer assumptions for flow in the region below the water table; that is, it assumes the vertical component of flow is negligible. This assumption leads to

$$K_m \frac{\partial}{\partial x} \left[(h + d) \frac{\partial h}{\partial x}\right] = \phi_e \frac{\partial h}{\partial t} \quad , \qquad 5.46$$

a result which is reasonably accurate if d is large compared to h and if the region of flow above the water table is small compared to d. Equation 5.46 is known as the Dupuit-Forchheimer equation.

Equation 5.46 is nonlinear because the coefficient of $\partial h/\partial x$ includes h. Glover linearized this equation by replacing h + d by an average value of h + d designated as $\bar{d}$. Therefore, Glover's equation has the form

$$\alpha \frac{\partial^2 h}{\partial x^2} = \frac{\partial h}{\partial t} \quad , \qquad 5.47$$

where

$$\alpha = \frac{K\bar{d}}{\phi_e} \quad .$$

This equation may provide satisfactory results (where Equation 5.46 is also valid) provided the slope of the water table is very small so that linearization does not introduce a significant error.

In many cases, however, $p_e/\rho g$ is not small compared to $\bar{d}$ so that it is not sufficient to neglect either flow or storage above the water table or to assume that the specific yield S_y is a constant equal to ϕ_e. Theoretically, the problem can be solved rigorously treating the system as a 2-phase flow region and solving the unsteady 2-phase flow Equation 5.3. For the 2-dimensional transient case under consideration, a rigorous solution requires a high-speed computer. The economics of agricultural drainage is usually such that the problem cannot be formulated in this way for routine design computations.

McWhorter and Duke (1976) have presented an approximate analytical solution which does not require a computer but which does account for flow and storage above the water table. Their solution also accounts (to a large extent) for the nonlinearity of the flow system.

A differential slice of soil, oriented parallel to the axis of the drains is shown in Figure 5-11. The depth below the soil surface to the top of the saturated region is denoted by z_d and the thickness of the fully saturated region (except for trapped air) above the water table is ψ_d. H_k is the effective permeable height used by Duke (1973) for the analysis of steady flow towards parallel drains [Section 4.4]. H_k is a height of saturated soil having the same capacity to transmit water as does the actual soil region above the water table. The mathematical expression for H_k is given by

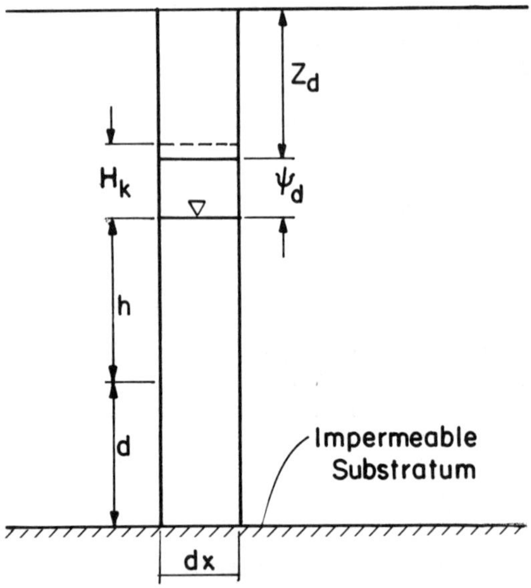

Figure 5-11. Differential slice of soil, oriented parallel to axis of drains [McWhorter and Duke (1976)].

$$H_k = \frac{1}{K_m} \int_0^H K_w \, dz \quad .$$

With this addition to the flow section, Equation 5.44 becomes

$$K_m \frac{\partial}{\partial x} [(h + d + H_k) \frac{\partial h}{\partial x}] = \frac{\partial V}{\partial t} \qquad 5.48$$

in which V is the volume of drainable water per unit surface area stored in the slice at any time.

An approximate expression for V is obtained using the assumption that the distribution of water in the slice is as if the vertical component of flow is negligible. This is a reasonable assumption since in field situations the rate of lowering of the water table is extremely slow. The empirical expression for effective water saturation of Brooks and Corey (1966),

$$S_e = (\frac{z_d - z + \psi_d}{\psi})^{-\lambda} \quad ,$$

is used to evaluate water content at any point above the fully saturated region. Integration of $\phi_e(1-S_e)dz$ from the top of the fully saturated region to the soil surface $(z = 0)$ and differentiation of the result with respect to time yields

$$\frac{\partial V}{\partial t} = \phi_e \psi_d [1 - (\hat{z}_d + 1)^{-\lambda}] \frac{\partial \hat{z}_d}{\partial t} \qquad 5.49$$

in which $\hat{z}_d = z_d/\psi_d$. Equation 5.49 is a tractable continuity equation which accounts for the major effects of storage above the water table.

Substituting Equation 5.49 into Equation 5.48, scaling all quantities having a dimension of length by ψ_d, and noting that $\hat{z}_d = D - h - 1$ results in

$$\frac{K_m}{\sigma_e \psi_d} \frac{\partial}{\partial \hat{x}} [(\hat{h} + \hat{d} + \hat{H}_k) \frac{\partial \hat{h}}{\partial \hat{x}}] = [1 - (\hat{D} - \hat{h})^{-\lambda}] \frac{\partial \hat{h}}{\partial t} \quad . \qquad 5.50$$

For a static water distribution, the Brooks-Corey relationship for $K_w(p_c)$ gives

$$\hat{H}_k = 1 + \frac{1}{1 + 3\lambda} [1 - (\frac{1}{\hat{D} - \hat{h}})^{1 + 3\lambda}] \quad .$$

In cases of practical interest (where the water table is not very near the soil surface), $\hat{h} \ll \hat{D}$, so that

$$\hat{H}_k \cong 1 + \frac{1}{1 + 3\lambda} [1 - (\frac{1}{\hat{D}})^{1 + 3\lambda}] \quad . \qquad 5.51$$

In this analysis, Equation 5.51 is used to evaluate $\hat{H}_k$. In most cases of practical interest $\hat{H}_k < 1.4$. Note that in this formulation $\hat{H}_k$ is treated as a constant, its dependence upon $\hat{h}$ being considered negligible.

Expanding the derivative on the left side of Equation 5.50 gives

$$\frac{\partial}{\partial \hat{x}} [(\hat{h} + \hat{d} + \hat{H}_k) \frac{\partial \hat{h}}{\partial \hat{x}}] = (\hat{h} + \hat{d} + \hat{H}_k) \frac{\partial^2 \hat{h}}{\partial \hat{x}^2} + (\frac{\partial \hat{h}}{\partial \hat{x}})^2 \quad .$$

Except in the immediate vicinity of the drains at small times, the second term on the right is negligible compared to the first term on the right, so that

$$\alpha(\hat{h}) \frac{\partial^2 \hat{h}}{\partial \hat{x}^2} = \frac{\partial \hat{h}}{\partial \hat{t}} \qquad 5.52$$

in which

$$\alpha(\hat{h}) = \frac{K_m}{\phi_e \psi_d} \left[\frac{\hat{h} + \hat{d} + \hat{H}_k}{1 - (\hat{D} - \hat{h})^{-\lambda}}\right] \quad .$$

Equation 5.52 retains the most important effects of the non-linearity of flow.

The conditions imposed on Equation 5.52 are expressed as

$$\hat{h}(0,\hat{t}) = \hat{h}(\hat{L},\hat{t}) = 0 \quad .$$

This condition ignores the existence of the seepage face at the drain wall. However, Equation 5.52 implies that the entire interval $\hat{H}_k$ is available for flow, including that part where the water is at sub-atmospheric pressure. The latter simplification should tend to compensate in some degree for ignoring the seepage face.

Duke (1973) obtained a finite difference solution for Equation 5.52. Later, McWhorter and Duke (1976) found an approximate analytical solution for this equation. Figure 5-12

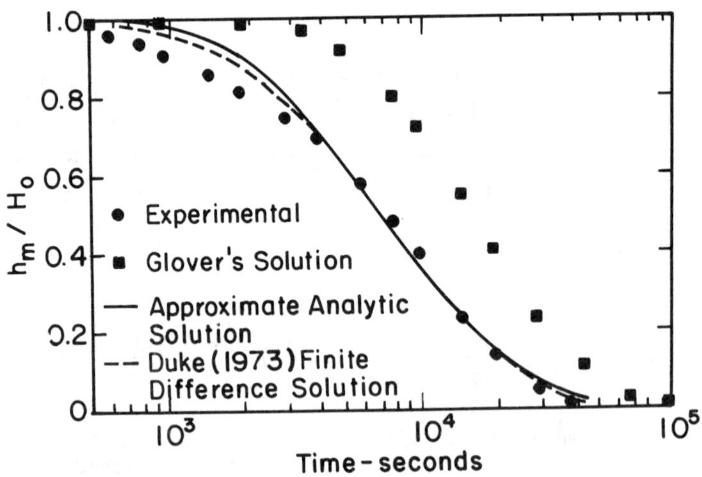

Figure 5-12. Drawdown at midpoint between parallel drains as a function of time [McWhorter and Duke (1976)].

shows some typical results from these solutions compared with experimental data obtained by Hedstrom et al. (1971) with a large soil flume. Results from the solution of Glover's linear Equation 5.47 also are shown in this figure. In Figure 5-12, h_m is the water table elevation above the drains at the mid-point and H_o is the initial water table elevation [see Figure 5-10]. Equation 5.52 clearly provides a better approximation to the measured data than does Equation 5.47 for this case.

5.3 DISPLACEMENT OF PETROLEUM FLUIDS

Petroleum fluids, e.g., crude oil and natural gas, are produced from porous rocks, usually aquifers containing brine, by displacement of one fluid by another. The displacement process may involve the displacement of oil by brine from surrounding rock or from injection wells, or it may involve the displacement of oil by natural gas evolving from solution or invading from an expanding gas cap. In such systems, brine is usually the wetting phase and oil is the nonwetting phase. When gas is present, it is the nonwetting phase in respect to either oil or brine. The displacement process may occur spontaneously when a well is drilled into the aquifer, or it may be induced by an artificial injection of either brine or gas.

In some cases, fluids may be produced from an aquifer simply by compression of the pore space upon the reduction of fluid pressure or by the expansion of the fluids. The matrix compression and fluid expansion are significant mainly when they take place over a very large volume of the aquifer, and the effect locally (at the site of a producing well) is mostly the replacement of petroleum fluids by fluids being "squeezed" out of the surrounding aquifer.

The theory used in analyzing petroleum displacements is basically the same as that for air-water displacements, since both involve the displacement of one <u>immiscible</u> fluid by another. Consequently, the starting equation is the same, i.e., Equation 5.3. However, the relative magnitudes of the several forces producing the flow are often quite different. Therefore, the simplifications made in the case of petroleum displacements are likely to be very different from those made for air-water displacements, as explained in Section 5.1.1.

5.3.1 Linear displacement - The displacement of one fluid by another in a 3-dimensional system is difficult to analyze mathematically. Fortunately, considerable insight into the practical problems encountered can be obtained by analyzing one-dimensional flow systems. Furthermore, many field situa-, tions can be approximated by assuming one-dimensional models as an idealization of the real situation (Richardson (1961)].

Most oil producing aquifers are thought to have been originally saturated with brine which subsequently was partly replaced by oil and perhaps gas, although brine remains as a residual wetting phase. The question to be answered is how much oil will be produced when a well is drilled into the formation before the operation of the well is no longer economical. The operation will become uneconomical some time after the invading brine breaks through to the well and the production of brine becomes very large compared to the production of oil. Often, the brine comes from a line of injection wells and the brine advances along a broad wetting front toward a line of producing wells.

Such a system may sometimes be idealized as a linear displacement process. A long thin tube of homogeneous porous material is considered as an example of such an idealization. The tube may be inclined at an angle β to a horizontal plane. The porous material is originally fully saturated with brine, but the brine (in excess of residual saturation) has been replaced later by oil. The tube of porous material is then in a condition called the restored state which is presumed to be similar to the condition of many petroleum aquifers before petroleum production begins.

Subsequently, some of the oil is replaced by brine flooding the tube from one end where it is introduced at high pressure. A question to be answered is how will the fluids be distributed along the tube at any time during the flooding process. Another question is how much oil will be produced from the outflow end of the tube as a function of time and as a function of the injected volume of brine.

The analysis of the idealized case provides some insights into the practical problem. The formulation of a flow equation for this case involves the simplification of Equation 5.3. Most petroleum aquifers have a small slope and the displacement takes place under very large pressure gradients over large distances where the fluids flow side by side with small saturation gradients. Therefore, it is reasonable to drop the first term on the left of Equation 5.3. Furthermore, the idealized case is one of linear displacement so that the directional subscripts also can be dropped. Consequently, an appropriate flow equation is

$$\phi \frac{\partial S}{\partial t} = - \frac{\partial}{\partial x} [q_{nw}(\frac{\mu_{nw} k_w}{\mu_w k_{nw}})] \quad . \qquad 5.53$$

Under the condition of a negligible capillary pressure gradient, the term on the right of Equation 5.53 is an approximation of $-\partial q_w/\partial x$. However, q_w varies with x so that q_w is a varying fraction F_w of the total flux rate q_t, i.e.,

$q_w + q_{nw}$. The fraction F_w is assumed to be a function of S only, for a given pair of fluids. Equation 5.53 is rewritten in terms of q_t and F_w as

$$\phi \frac{\partial S}{\partial t} = - \frac{\partial}{\partial x} (q_t F_w) \quad .$$

The assumption that the fluids are incompressible is equivalent to assuming that q_t does not vary with x so that

$$\phi \frac{\partial S}{\partial t} = - q_t \frac{\partial F_w}{\partial x} \quad .$$

Since F_w is presumed to be a function of S only,

$$\frac{\partial S}{\partial t} = - (\frac{q_t}{\phi} \frac{dF_w}{dS}) \frac{\partial S}{\partial x} \quad . \qquad 5.54$$

The saturation S is a function of x and t only, where x is regarded as the coordinate of a plane (perpendicular to the tube) where S has any specified value. Therefore, the total derivative of S with respect to time is given by

$$\frac{dS}{dt} = \frac{\partial S}{\partial x} \frac{dx}{dt} + \frac{\partial S}{\partial t} \quad .$$

The quantity dx/dt is interpreted as the velocity of a plane (perpendicular to the tube) having any specified value of S, i.e., $dx/dt = (dx/dt)_S$. Since $(dS/dt)_S = 0$,

$$(\frac{dx}{dt})_S = - \frac{\partial S}{\partial t} / \frac{\partial S}{\partial x} \quad .$$

Combining this with Equation 5.54 gives

$$(\frac{dx}{dt})_S = \frac{q_t}{\phi} (\frac{dF_w}{dS})_S \quad , \qquad 5.55$$

which is called the Buckley-Leverett equation [Buckley, S. E. and Leverett, M. C. (1942)].

If the permeability ratio k_{nw}/k_w as a function of S is known, dF_w/dS can be evaluated as a function of S. When S is known at t = 0, Equation 5.55 can be integrated to give the saturation distribution for any t > 0. Unfortunately, particular values of dF_w/dS may occur at two different values of

S, so that two different values of S are calculated for particular values of $(x)_S$, a physical impossibility. This problem is resolved by postulating a discontinuity in S at a front, and requiring that a material balance be satisfied [see Section 5.2.7]. An explanation of this procedure is presented in a paper by Buckley and Leverett (1942) and in texts on petroleum engineering, including one by Collins (1961). However, the above procedure can give only a rough approximation of the real distribution of fluids near the front because the effect of capillarity is substantial in this region and this effect is neglected in Equation 5.55.

The integration of Equation 5.55 with respect to time gives

$$(x)_S = \frac{Q_i(t)}{\phi A} \left(\frac{dF_w}{dS}\right)_S \qquad 5.56$$

in which $Q_i(t)$ is the cumulative total volume of fluid flowing through an area A in an interval of time t, and $(x)_S$ is the distance traveled along x by the point of saturation S in the same interval of time. Experience has shown that Equation 5.56 gives a good approximation of the actual distribution of S for long systems with large flow rates, except near the advancing front as previously noted.

5.3.2 Determination of permeability ratios using the Welge technique - When conditions are such as to justify the use of Equation 5.55, it can be integrated using a method proposed by Welge (1952).

The following analysis applies to the displacement of oil by brine from a horizontal tube of homogeneous porous material which has been brought to a restored state as described in Section 5.3.1. In this case, the tube has a length L, area A and a zero angle of inclination. At time t = 0, brine is introduced at x = 0 at very high pressure. This causes brine to flood the tube at a rate q_w which is a function of time.

At x = 0, only brine is flowing at t > 0, while at x = L, only oil flows initially. After breakthrough of brine, both fluids flow at x = L. Designating the cumulative inflow by Q_i, the cumulative outflow of oil by Q_o (and brine by Q_w), it is noted that $Q_i = Q_o + Q_w$.

At the outflow end,

$$F_o = \frac{dQ_o}{dQ_i} \quad ,$$

where F_o is the fraction of the total volume flux that consists of oil flux. For the conditions under consideration, $F_o \approx f_{nw}$, where f_{nw} is an approximation of F_{nw} derived as follows:

An assumption made for the derivation of the Buckley-Leverett equation is that both fluids flow under nearly the same potential gradient. Therefore,

$$q_{nw} \approx - \frac{k_{nw}}{\mu_{nw}} \left(\frac{\partial p^*}{\partial x}\right)$$

and

$$q_t \approx - \left(\frac{k_{nw}}{\mu_{nw}} + \frac{k_w}{\mu_w}\right)\left(\frac{\partial p^*}{\partial x}\right) \quad ,$$

where p^* is the piezometric potential. Consequently,

$$\frac{q_{nw}}{q_t} \approx \left(1 + \frac{\mu_{nw}\, k_w}{\mu_w\, k_{nw}}\right)^{-1} = f_{nw} \quad .$$

An analogous approximation of F_w is designated as f_w.

The approximation for F_o leads to

$$\left(1 + \frac{\mu_{nw}\, k_w}{\mu_w\, k_{nw}}\right)^{-1} \approx \frac{dQ_o}{dQ_i} \quad .$$

Solving the latter equation explicitly for the permeability ratio k_w/k_{nw} gives

$$\frac{k_w}{k_o} = \frac{\mu_w}{\mu_o} \left(\frac{dQ_i}{dQ_o} - 1\right) \quad . \qquad 5.57$$

Consequently, the permeability ratio at any $t > 0$ can be determined from the ratio of brine inflow to oil outflow rates. The ratio is zero until brine breakthrough after which it gradually increases and eventually becomes infinitely large when the outflow ceases completely.

The value of S at the outflow end of the tube can be computed as a function of Q_i and dQ_i/dQ_o by solving a material balance equation; i.e.,

$$\phi \int_0^L (1 - S)dx = \phi L(1 - S_r) - (Q_o/A) \ ,$$

in which A is the cross-sectional area of the tube and L is its length. Integrating by parts gives

$$\phi L(1 - S_L) - \phi \int_{S_o}^{S_L} x(-dS) = \phi L(1 - S_r) - (Q_o/A) \quad ,$$

where S_L is S at x = L. Note that (-dS) is inserted in the integral because this is equivalent to d(1-S). The equation may be rewritten as

$$S_L = S_r + \frac{Q_o}{A\phi L} + \frac{1}{L}\int_{S_o}^{S_L} x \, dS \quad . \qquad 5.58$$

It was pointed out by Welge (1952) that the sum of the first two terms on the right of this equation gives the average wetting phase saturation of the tube at any time during the flood.

According to Equation 5.56, the third term on the right of the expression for S_L can be written as

$$\frac{Q_i}{A\phi L} \int_{x=0}^{x=L} dF_w$$

or

$$- \frac{Q_i}{A\phi L} \int_{x=0}^{x=L} df_{nw} \quad .$$

At the inflow end, $f_{nw} = 0$, and at the outflow end, $f_{nw} = dQ_o/dQ_i$. Therefore,

$$S_L = S_r + \frac{Q_o}{A\phi L} - \frac{Q_i}{A\phi L} \left(\frac{dQ_o}{dQ_i}\right)_L \quad . \qquad 5.59$$

All quantities on the right of Equation 5.59 can be determined from inflow and outflow data as is the case for Equation 5.57.

Equation 5.57 and 5.59 provide a method of determining permeability ratios as a function of S on small rock cores. The method is widely used in the petroleum industry. Cores of rock from producing formations are obtained with a core drill at the time wells are drilled. Such cores are usually about 10 cm in diameter. In order to obtain cores parallel to the bedding, smaller cores are removed from the well cores with the axis of the smaller cores being more or less parallel to the bedding. The latter cores are usually about 2 to 2.5 cm in diameter and about 5 to 8 cm in length.

The residual fluids are extracted from the laboratory cores and the porosities are determined. The surface of the cores (except for the ends) are sealed, usually with an acrylic plastic. The cores are then saturated with brine. Next, all of the brine except residual brine is replaced by oil. The core is then in the restored state as previously described.

Finally, brine is injected into one end of the core at a very high pressure and the outflow of both oil and brine from the opposite end are measured separately until the outflow of oil nearly stops. The cumulative outflow of oil Q_o is plotted as a function of the total outflow Q_i. The slope of the curve, dQ_o/dQ_i, is substituted into Equations 5.57 and 5.59 to obtain k_w/k_o as a function of S.

Equations 5.57 and 5.59 apply strictly for long tubes only. However, it is believed that the error introduced by their application to short cores is minimized by using larger pressure gradients. The Buckley-Leverett equation also is used to describe displacement of oil by gas from short cores. In this case, where the displacing fluid is a gas, the error is likely to be more serious. However, the error is thought to be substantially reduced by first saturating the core with an oil of very high viscosity which permits displacement experiments to be carried out under an extremely large pressure gradient. In any case, this technique is widely used in the petroleum industry to obtain k_g/k_o ratios as a function of S. The ratios k_g/k_o and k_o/k_w are used in the petroleum industry to predict reservoir behavior as explained in Section 5.3.5. Both of these ratios tend to plot as a straight line on a semi-log plot of $k_{nw}/k_w(S)$, as shown in Figure 5-13.

5.3.3 Calculation of relative permeabilities from linear displacement experiments - For some calculations, the determination of k_{nw}/k_w curves is not sufficient and individual relative permeability curves for each flowing fluid are needed as well. Johnson et al. (1959) have shown that linear displacement experiments on small cores can be used to determine k_{rnw} and k_{rw} as functions of S.

The analysis upon which the method of Johnson et al. is based makes use of several relationships that can be deduced from the Buckley-Leverett and Welge theories. For a water-oil system, these are:

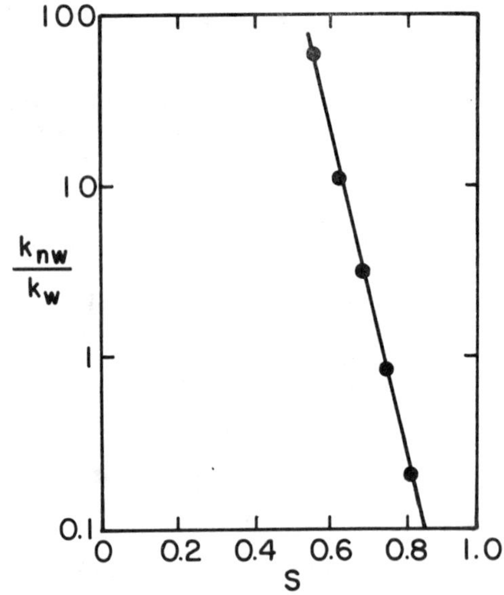

Figure 5-13. Permeability ratio as a function of saturation.

(1) $f_o\, q_t = - \frac{k_o}{\mu_o} \frac{\delta p}{\delta x}$, which is Darcy's equation for the oil phase.

(2) $\frac{f_w}{f_o} = \frac{k_w\, \mu_o}{k_o\, \mu_w}$, which can be deduced from the definitions of f_w and f_{nw}.

(3) $S_{av} = (S)_L + \frac{Q_i}{A\phi L} (f_o)_L$, where the subscript L indicates that the quantities are evaluated at the outflow end of the sample. This is the Welge Equation 5.58.

(4) $(\frac{dx}{dt})_S = \frac{q_t}{\phi} (\frac{dF_w}{dS})_S$, the Buckley-Leverett equation.

(5) $(x)_S = \frac{Q_i}{\phi A} (\frac{dF_w}{dS})_S$, obtained by integrating the Buckley-Leverett equation over an interval of time and distance of travel of the saturation S.

(6) $\frac{Q_i}{A\phi L} = \frac{1}{(df_w/dS)_L}$, which comes from evaluating the relationship (5) at the outflow end where $(x)_S = L$. Note that $Q_i/A\phi L$ represents the number of pore volumes injected.

(7) $(f_o)_L = A\phi L \frac{dS_{av}}{dQ_i}$, which is obtained by differentiating the relationship (3) with respect to Q_i. Note that for a fixed value of $(f_o)_L$, S_L is a constant.

The sample preparation and experimental procedure is the same as for the determination of k_o/k_w curves described in Section 5.3.2.

At a given instant during the displacement experiment, the pressure drop across a sample of length L is given by

$$\Delta p = \int_0^L \frac{\delta p}{\delta x}\, dx \quad ,$$

or from (1)

$$\Delta p = - q_t \mu_o \int_0^L \frac{f_o}{k_o}\, dx \quad . \qquad 5.60$$

According to (5) and (6), the various saturations (at a particular instant) occur at distances which can be calculated from

$$\frac{(x)_S}{L} = \frac{(df_w/dS)_S}{(df_w/dS)_L} \quad ,$$

or

$$x = \frac{(f'_w)_S}{(f'_w)_L}\, L \quad ,$$

and

$$dx = \frac{L}{(f'_w)_L}\, df'_w \quad ,$$

in which $(f'_w)_L$ has some fixed value depending upon the particular instant under consideration. Substituting this expression for dx into Equation 5.60 gives

$$\frac{\Delta p (f'_w)_L}{q_t \mu_o L} = \int_0^{(f'_w)_L} \frac{f_o}{k_o}\, df'_w \quad .$$

Substituting the product $k_{ro}\, k_m$ for k_o gives

$$\frac{\Delta p k_m (f'_w)_L}{q_t \mu_o L} = \int_0^{(f'_w)_L} \left(\frac{f_o}{k_{ro}}\right) df'_w \quad . \qquad 5.61$$

According to Johnson et al. (1959), the quantity $q_t \mu_o L/\Delta p k_m$ has been called relative injectivity I_r by Rapoport. I_r and $(f'_w)_L$ are interrelated since both Δp and $(f'_w)_L$ vary with the amount of water injected. Differentiating both sides of Equation 5.61 with respect to $(f'_w)_L$ results in

$$\frac{d[(f'_w)_L/I_r]}{d(f'_w)_L} = \left(\frac{f_o}{k_{ro}}\right)_L \quad .$$

Substituting the relationship (6) into the latter expression gives

$$\frac{d\left(\frac{1}{Q_i I_r}\right)}{d\left(\frac{1}{Q_i}\right)} = \left(\frac{f_o}{k_{ro}}\right)_L \quad . \qquad 5.62$$

During the experiment, Q_i and Q_o are recorded as well as Δp. This permits the evaluation of the derivatives in Equation 5.62 and in the relationship (7). From (7), $(f_o)_L$ is determined which permits the evaluation of $(k_{ro})_L$. The corresponding value of $(S)_L$ is determined from (3), S_{av} being given by

$$S_{av} = S_r + \frac{Q_o}{A\phi L} \quad .$$

5.3.4 Calculation of relative permeabilities by application of linear scaling principle - A displacement resulting from the injection of a single fluid is said to be linearly scalable if the average saturation of any portion of a porous sample is a function only of the number of pore volumes of displacing fluid that have been injected into or through the portion under consideration [Parsons, R. W. and Jones, S. C. (1977)]. The processes discussed in Section 5.3 are linearly scalable if the assumptions made concerning these processes are valid. In fact, Equation 5.56 implies that the displacement is linearly scalable.

Jones and Roszelle (1978) have shown how the acceptance of the principle of linear scalability can simplify the mathematics and expedite the calculations involved in the determination of relative permeabilities from displacement experiments on laboratory samples. To understand their procedures, it is necessary first to examine the concept involved in specifying the <u>number of pore volumes injected</u>.

The concept can be examined in reference to Figure 5-14.

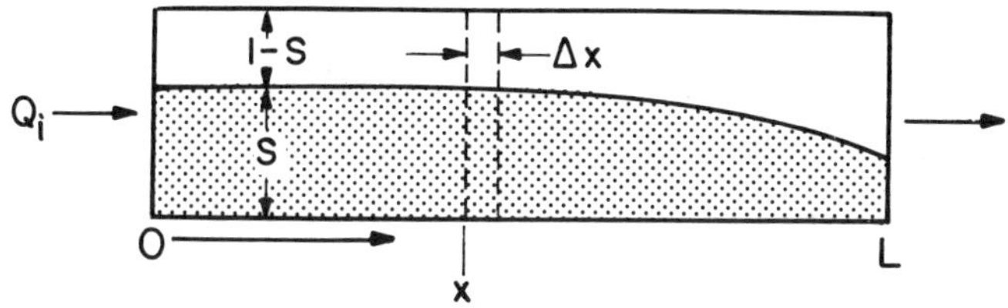

Figure 5-14. Schematic of sample undergoing linear displacement of oil by water.

In Figure 5-14, Q_i indicates a volume of water that has been injected through an area A at time t (through the input face of a linear sample initially saturated with oil and residual water). The height of the shaded area indicates the water saturation S at any vertical plane along the length of the sample. The sample has a length L, and the coordinate x indicates a distance along the sample beginning at x = 0. The number of pore volumes injected PV_i for the entire sample at any time is $Q_i/\phi AL$ whereas PV_i for the portion extending from 0 to x is a larger number given by $Q_i/\phi Ax$. Note that as $x \rightarrow 0$, $PV_i \rightarrow \infty$, even for a small value of Q_i.

The water saturation increases and the oil saturation decreases as PV_i increases. The value of S reaches some limiting value when $PV_i \rightarrow \infty$. Theoretically, the limiting (maximum) value of S is obtained instantly (upon the beginning of injection) at the input surface where x = 0. The value of S decreases with x and is smallest at x = L. However, when Q_i becomes very large, the variation of S with x is very small, and as $Q \rightarrow \infty$, S is theoretically at the limiting value at all x.

A mathematical advantage is gained by starting with the assumption that:

(1) Saturation and other properties related to S are single-valued functions of PV_i only.

(2) The same functional relationship between any of these properties and PV_i holds for values of the variables averaged over a segment of a homogeneous sample regardless of the size of the segment, provided the segment is large enough to be a representative element of the porous medium.

It is possible to show that any property satisfying the condition of being a single-valued function of S will also be a single-valued function of PV_i only. In fact, the theory of Buckley and Leverett (1942) and Welge (1952) imply this is the case provided the relative permeabilities are single-valued functions of S.

To demonstrate the mathematical consequences of the linear scaling principle, an unspecified property p is considered which satisfies the linear scaling conditions. This property might be S or some function of S, e.g., f_w, k_{rw} or k_{ro}. The following analysis also can be applied to any function of the latter variables, for example, it also applies to p^{-1}. In general, p varies with x and Q_i in a cylindrical porous sample during a displacement process in which the displacing fluid is injected through a face perpendicular to x at $x = 0$. The value of p at x for a particular Q_i is defined by

$$p_x = \lim_{\Delta x \to 0} \frac{(x + \Delta x)\, \bar{p}_{(x+\Delta x)} - x\, \bar{p}_x}{\Delta x}$$

or

$$p_x = \lim_{\Delta x \to 0} \left[\bar{p}_{(x+\Delta x)} + \frac{x(\bar{p}_{(x+\Delta x)} - \bar{p}_x)}{\Delta x}\right]$$

in which $\bar{p}_x$ is the average value of over the segment extending from 0 to x. Therefore, from the definition of a derivative

$$p_x = \bar{p}_x + x \frac{\partial \bar{p}_x}{\partial x} ,$$

in which all quantities including the derivative are functions of PV_i. Since $\bar{p}_x$ is assumed to be a single-valued function of PV_i only, this also can be written as

$$p_x = \bar{p}_x + x \left[\frac{d\bar{p}_x}{d(PV_i)} \frac{\partial (PV_i)}{\partial x}\right]_x .$$

Because PV_i is $Q_i/\phi Ax$,

$$\frac{\partial(PV_i)}{\partial x} = -\frac{Q_i}{\phi Ax^2}$$

so that

$$p_x = \bar{p}_x - (PV_i)_x \left[\frac{d\bar{p}_x}{d(PV_i)}\right]_x . \qquad 5.63$$

Applying Equation 5.63 to S_x at $x = L$, and noting that the increase in S_{av} of the sample as a whole is related to the outflow of oil Q_o, gives

$$S_L = S_{av} - \frac{Q_i}{\phi AL}\left(\frac{dQ_o}{dQ_i}\right)_L ,$$

which is Equation 5.59. Jones and Roszelle (1978) have presented a convenient graphical procedure for evaluating S_L using plots of S_{av} as a function of Q_i from laboratory displacement data.

An expression for $(k_{rw})_L$ corresponding to S_L also can be derived using the linear scaling principle. The procedure employs the concept of an <u>effective viscosity</u> (or <u>relative reciprocal mobility</u> Λ_r^{-1}) which is defined so that a flow equation for q_t is

$$q_t = -\frac{k}{\Lambda_r^{-1}}\left(\frac{dp}{dx}\right)_x \qquad 5.64$$

in which k is the permeability when the sample is fully saturated with water and $(dp/dx)_x$ is assumed to apply for either the water or the oil phase. By writing separate equations for q_w and q_o and adding these to obtain q_t, it can be shown that

$$\Lambda_r = \left(\frac{k_{rw}}{\mu_w} + \frac{k_{ro}}{\mu_o}\right) = \Lambda/k , \qquad 5.65$$

which is called <u>relative mobility</u>, Λ being called <u>total mobility</u>. A second equation for q_t is given by

$$(f_w)_x\, q_t = -\frac{k\, k_{rw}}{\mu_w}\left(\frac{dP}{dx}\right)_x .$$

Equating the latter expression to Equation 5.65 and solving for k_{rw} at the outflow end gives

$$(k_{rw})_L = (f_w)_L \mu_w (\Lambda_r)_L \quad . \qquad 5.66$$

Since k_{rw} and k_{ro} are considered to be single-valued functions of S, it follows that Λ_r and Λ_r^{-1} are also single-valued functions of PV_i. Therefore, Equation 5.63 can be applied to Λ_r^{-1}, i.e.,

$$(\Lambda_r^{-1})_L = (\Lambda_r^{-1})_{av} - \frac{Q_i}{\phi AL} \left[\frac{d(\Lambda_r^{-1})_{av}}{d(PV_i)}\right]_L \quad . \qquad 5.67$$

Jones and Roszelle (1978) have described a graphical procedure for solving Equation 5.67 similar to that used by them to obtain S_L. However, the data needed are values of $(\Lambda_r^{-1})_{av}$ as a function of PV_i. Values of $(\Lambda_r^{-1})_{av}$ are obtained as a function of PV_i from the experimental data which include the Δp across the entire sample; i.e.,

$$(\Lambda_r^{-1})_{av} = \frac{k}{q_t} \frac{\Delta p}{L} \qquad 5.68$$

which results from integrating Equation 5.64 over x from 0 to L.

The linear scaling principle and the procedures employed by Jones and Roszelle involve the same theoretical limitations as the procedures for Johnson et al. (1959). These include the requirement that $|dP/dx|$ is very large compared to $|dP_c/dx|$ and that k_{rw} and k_{ro} are single-valued functions of S only. The latter assumption may not be entirely valid since there is evidence that both k_{rw} and k_{ro} are to some extent dependent on q_t. This has been observed particularly in respect to the residual oil saturation during water floods, the residual oil saturation supposedly representing the saturation at which k_{ro} is zero. However, increasing the injection rate decreases the residual oil saturation slightly. Lowering the oil-water interfacial tension also reduces the residual oil saturation. Furthermore, injection rates necessary to minimize the effect of dp_c/dx (in small laboratory samples) are much greater than those found in field applications. It is believed, however, that this difficulty is not very serious because there is evidence that k_{rw} and k_{ro} are not very sensitive functions or rate.

5.3.5 Prediction of reservoir behavior from permeability ratios

- Permeability ratios and relative permeabilities as functions of saturation constitute the basic data used in the prediction of petroleum reservoir behavior. The two-phase flow equation, i.e., Equation 5.3, provides the theory upon which the calculations are made.

The processes to be evaluated may involve water displacing either oil or gas (or a gas displacing oil). Sometimes more than one of these processes occur simultaneously. The displacements are rarely truly linear, but they often can be idealized as such with an error that is not large compared to other uncertainties involved.

Linear displacements are often described by the steady two-phase flow equation, i.e., Equation 4.2. However, it is convenient for many purposes to rearrange this equation and to write it explicitly for the ratio q_w/q_t, i.e., F_w. The result is

$$F_w = \frac{1 + \frac{k_{nw}}{q_t\ \mu_{nw}} \left(\frac{\partial p_c}{\partial x} + \Delta\rho g_i\right)}{1 + \frac{k_{nw}\ \mu_w}{k_w\ \mu_{nw}}} \quad . \qquad 5.69$$

Equation 5.69 is called the "Fractional Flow" equation. When both $\partial p_c/\partial x$ and $\Delta\rho g_i$ are negligible, $F_w \approx f_w$ as noted in Section 5.3.1. The latter situation frequently prevails for petroleum displacements, especially where brine displaces oil. However, it may sometimes be desirable to retain the term $\Delta\rho g_i$ where oil is displaced by gas. The retention of $\partial p_c/\partial x$ is rarely required in routine petroleum engineering calculations because field systems are usually long enough that this term is negligible.

In most cases which can be idealized as linear, Equation 5.56 (or a modification of it) can be used to predict reservoir behavior. For a displacement by water (brine), for example, it is often useful to predict the amount of oil that can be recovered before the water-oil ratio at the producing well becomes such that additional production is uneconomical.

The data needed, in addition to k_o/k_w curves, include the thickness of the oil-bearing stratum, the distance from the wetting front to the producing wells, the values of S_r, ϕ, k_m,

μ_w, μ_o and the water injection rate Q_i/t per unit width of stratum. If the term $\Delta\rho g_i$ is to be included in the expression for F_w, the densities ρ_o and ρ_w and the angle of inclination of the formation must also be included.

The first step in the computation is the determination of $F_w(S)$, which for the case of small $\Delta\rho$ and/or small angles of inclination is approximated by f_w, i.e., $[1 + (k_o\ \mu_w/k_w\ \mu_o)]^{-1}$. A plot of $f_w(S)$ is made and the slope df_w/dS is measured at intervals of S (say at increments of 0.05), beginning with some value of S greater than S_r. Typical $f_w(S)$ and $df_w/dS(S)$ functions are shown in Figure 5-15.

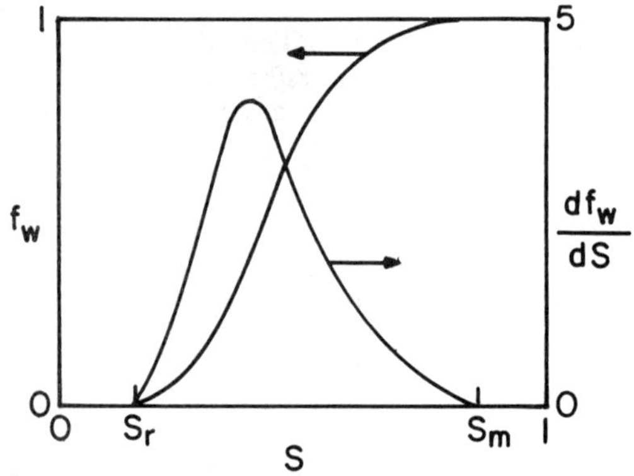

Figure 5-15. The fractional flow of wetting fluid and its derivative as a function of saturation.

Values of df_w/dS are recorded for the selected values of S. The distances travelled by the various saturations during a given time interval are determined from Equation 5.56, approximated by

$$(x)_S = \frac{Q_i}{\phi A} \left(\frac{df_w}{dS}\right)_S \quad ,$$

in which Q_i is the volume (per unit width of stratum) of brine injected during the same interval, and A is the thickness of the stratum.

A plot of $S(x)$ is made, as illustrated in Figure 5-8, for a number of time intervals. The curve identified as t_1 indicates that, in this time interval, the flood front has not yet reached the producing wells. Note, that before the time of breakthrough, the area under the curve (above S_r) must equal

$Q_i/\phi LA$; i.e., the volume of oil displaced must equal the volume of brine injected. After breakthrough, the area under the curve equals the volume of oil displaced but, in general, this will be less than the volume of brine injected because the wells will produce brine as well as oil.

Equation 5.56 does not permit an explicit solution of the maximum distance travelled by the flood front. In fact, the location of a given calculated saturation is double-valued, because df_w/dS is double-valued as explained in Section 5.3.1. The dotted line in Figure 5-8 illustrates this situation. However, the real saturation is not double-valued since a portion of the curve is not applicable to the displacement process. A discontinuity in saturation at the front is postulated such that the area under the curve (above S_r) behind the discontinuity equals $Q_i/A\phi L$. In an actual case, the saturation is not quite discontinuous at the front, the apparent discontinuity being a consequence of neglecting $\partial p_c/\partial x$ in the expression for F_w.

The Q_i for breakthrough can be determined by trial from plots as shown in Figure 5-16, and if Q_i/t is known, the time for breakthrough is also determined. After breakthrough, the

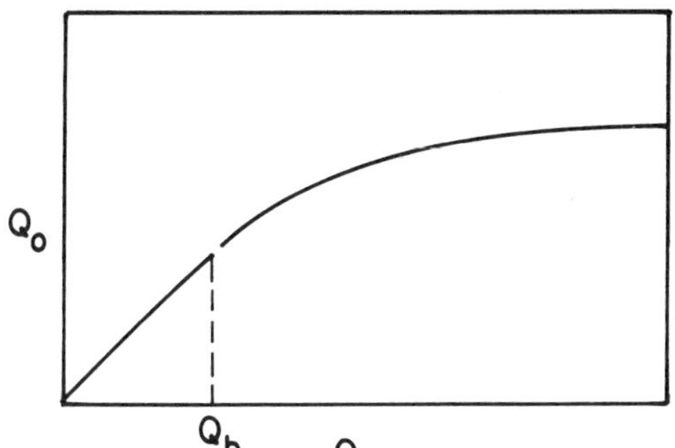

Figure 5-16. Oil recovered as a function of water injected for a linear displacement.

value of F_w gradually increases at the producing wells where $x = L$. The amount of oil that can be recovered by the time F_w reaches a specified value can be determined from a version of Equation 5.69 in the form

$$(1 - F_w)_L = \frac{Q_o/A\phi L - (S)_L + S_r}{Q_i/A\phi L} \quad . \qquad 5.70$$

As an example, consider the curve designated as t_2 in Figure 5-8. This curve has been plotted for a given time interval corresponding to a specified Q_i and a specified value of $(S)_L$. The value of $Q_o/A\phi L$ is given by the area under the curve above S_r. The specified values of Q_i and $(S)_L$ along with the computed value of $Q_o/A\phi L$ permits the calculation of F_w for this particular value of t and Q_i. The process is repeated for a number of values of Q_i. A plot of Q_o as a function of Q_i typically has characteristics as illustrated in Figure 5-16, in which Q_b indicates the water injected at the time of breakthrough.

Since the ultimate recovery of oil is approached asymptotically, it is never economically feasible to reach the ultimate recovery. The amount of water that must be produced becomes excessive before the ultimate recovery is achieved.

REFERENCES

Ames, W. F. (1965). Nonlinear Partial Differential Equations in Engineering, Academic Press, New York, N.Y., Vol. 18, Chapter 5, pp. 195-267.

Anat, A., Duke, H. R., and Corey, A. T. (1965). Steady upward flow from water tables. Colorado State University Hydrology Paper No. 7, June, 34 p.

Bear, J. (1972), Dynamics of Fluids in Porous Media. American Elsevier Publishing Co., Inc., New York, N.Y., p. 499.

Brooks, R. H. and Corey, A. T. (1966). Properties of porous media affecting fluid flow. ASCE, J. Irrig. Drain. Div., IR2, Vol. 92, pp. 61-88.

Bruce, R. R. and Klute (1956). The measurement of soil moisture diffusivity. SSSA Proceedings, Vol. 20, No. 4, pp. 459-462.

Buckley, S. E. and Leverett, M. C. (1942). Mechanism of fluid displacement in sands. Trans. AIME, Vol. 146, p. 107.

Collins, R. E. (1961). Flow of Fluids through Porous Materials. Reinhold Publishing Corporation, New York, Section 6.20, pp. 142-149.

Davidson, J. M., Nielsen, D. R. and Biggar, G. W. (1963). The measurement and description of water flow through Columbia Silt Loam and Hesperia Sandy Loam. Hilgardia, Vol. 34, pp. 601-617.

Dixon, R. M. and Linden, D. R. (1972). Soil-air pressure and water infiltration under border irrigation. SSSA Proceedings, Vol. 36, No. 5, pp. 948-953.

Duke, H. R. (1973). Drainage design based upon aeration. Colorado State University Hydrology Paper No. 61, June, 59 p.

Dumm, L. D. (1964). Transient-flow concept in subsurface drainage. Trans. ASAE, Vol. 7, pp. 142-146.

Dumm, L. D. (1967). The transient flow theory and its use in subsurface drainage of irrigated land. ASCE, Irrig. Drain. Div., Water Resources Conference, New York, N.Y., 38 p.

Gardner, W. R. (1962). Approximate solutions of a nonsteady-state drainage problem. SSSA Proceedings, Vol. 26, pp. 129-132.

Green, W. H. and Ampt, C. A. (1911). Studies on soil physics I. Flow of air and water through soils. Jour. Agr. Sci., Vol. 4, May, pp. 1-24.

Hanks, R. J. and Bowers, S. A. (1962). Numerical solution of moisture flow equation for infiltration into layered soils. SSSA Proceedings, Vol. 26, pp. 530-534.

Hedstrom, W. E., Corey, A. T., and Duke, H. R. (1971). Models for subsurface drainage. Colorado State University, Hydrology Paper No. 48, April, 56 p.

Jensen, M. E. and Hanks, J. R. (1967). Nonsteady-state drainage from porous media. J. of Irrig. and Drain. Div., Proceedings ASCE, IR3, September, pp. 209-231.

Johnson, E. F., Gassler, D. P., and Nauman, Y. O. (1959). Calculation of relative permeability from displacement experiments. Petroleum Trans., AIME, Vol. 216, p. 370.

Jones, S. C., and Roszelle, W. O. (1978). Graphical techniques for determining relative permeability from displacement experiments. Journal of Petroleum Technology, Vol. XXX, May, pp. 807-817.

King, L. G. (1964). Imbibition of fluids by porous solids. A Ph.D. dissertation, Colorado State University, Fort Collins, Colorado, September, 231 p.

Ligon, J. T., Johnson, H. P. and Kirkham, D. (1962). Unsteady-state drainage of fluid from a vertical column of porous material. J. of Geophysical Research, Vol. 16, pp. 5199-5204.

McWhorter, D. B. (1971). Infiltration affected by flow if air. Colorado State University Hydrology Paper No. 49, May, Fort Collins, Colorado, 43 p.

McWhorter, D. B. (1976). Vertical flow of air and water with a flux boundary condition. Trans. ASAE, Vol. 19, No. 2, pp. 259-261.

McWhorter, D. B. and Duke, H. R. (1976). Transient drainage with non-linearity and capillarity. ASCE, J. Irrig. Drain. Div., IR-2, Vol. 102, No. 2, pp. 193-204.

Mein, R. G. and Larson, C. L. (1973). Modeling infiltration during a steady rain. Water Resour. Res. Jour., Vol. 9, No. 2, April, pp. 384-394.

Miller, E. E. and Elrick, D. E. (1958). Dynamic determination of capillary conductivity extended for non-negligible membrane impedance. SSSA Proceedings, Vol. 22, pp. 483-486.

Morel-Seytoux, H. J. (1973). Two-phase flows in porous media. Advances in Hydroscience, Vol. 9, pp. 119-202.

Morel-Seytoux, H. J. (1975). Derivation of equations for rainfall infiltration. CEP75-76HJM7, ERC, Colorado State University, Fort Collins, Colorado 80523, 26 p.

Morel-Seytoux, H. J. and Khanji (1974). Derivation of an equation of infiltration. Water Resour. Res. Jour., Vol. 10, No. 4, August, pp. 795-800.

Parlange, J. Y. (1972). Theory of water movement in soils: 8. One-dimensional infiltration with constant flux at the surface. Soil Science, Vol. 114, No. 1, pp. 1-4.

Parsons, R. W. and Jones, S. C. (1977). Linear Scaling in Slug-Type Processes - Application to Micellar Flooding. Society of Petroleum Engineers Journal, Vol. 17, No. 1, pp. 11-26.

Peck, A. J. (1965). Moisture profile development and air compression during water uptake by bounded porous bodies, 3: vertical columns. Soil Science, Vol. 100, No. 1, pp. 44-51.

Philip, J. R. (1957). The theory of infiltration: 1. The infiltration equation and its solution. Soil Science, Vol. 83, pp. 345-357.

Richards, L. A. (1931). Capillary conduction of liquids through porous mediums. Physics, Vol. 1, pp. 318-333.

Richardson, J. G. (1961). Flow through porous media. Section 16, Handbook of Fluid Dynamics, edited by V. I. Streeter, McGraw-Hill Book Co., Inc., New York, N.Y, pp. 16-3 through 16-112.

Rubin, J. and Steinhardt, R. (1963). Soil water relations during rain infiltration: 1. Theory. SSSA Proceedings, Vol. 27, No. 3, pp. 246-251.

Smith, R. E. (1972). The infiltration envelope: Results from a theoretical infiltrometer. Jour. of Hydrology, Vol. 117, pp. 1-21.

Welge, H. J. (1952). A simplified method for computing oil recovery by gas or water drive. Trans. AIME, Vol. 195, pp. 91-98.

Whisler, F. D. and Bouwer, H. (1970). Comparison of methods for calculating vertical drainage and infiltration for soils. J. Hydrology, Vol. 10, No. 1, pp. 1-19.

Youngs, E. G. (1960). The drainage of liquids from porous materials. J. of Geophysical Research, Vol. 65, pp. 4025-4030.

Youngs, E. G. and Peck, A. J. (1964). Moisture profile development and air compression during water uptake by bounded porous bodies: 1. Theoretical introduction. Soil Science, Vol. 98, pp. 280-294.

PROBLEMS AND STUDY QUESTIONS

1. When formulating problems involving displacement of one fluid by another, petroleum reservoir engineers frequently state that they have "ignored capillary effects." Explain in what sense they have, in fact, ignored capillary effects and under what conditions this is a valid approximation.

2. The integrated form of the Buckley-Leverett equation (with the term $\partial p_c/\partial x$ omitted) gives a reasonably accurate approximation of the actual distribution of S for long systems with high flow rates. Explain why the length of a system is significant in this respect.

3. What difficulties would be expected in using the technique of Johnson et al. (based on the Welge integration of the Buckley-Leverett equation) for determining air and water relative permeabilities of soil materials.

4. Starting with the 2-phase unsteady flow equation, show that the fact (of air viscosity being much smaller than water viscosity) is not sufficient to justify neglecting air resistance in a water-air displacement process.

5. In a soil water-air system, it is reasonable to assume that the fluids are incompressible, and that

$$\frac{\partial q_t}{\partial x_i} = 0.$$

   In view of this fact, explain why it is not (in general) valid to drop the last term from the 2-phase unsteady flow equation.

6. Explain why the "diffusivity" equation is not valid for the description of flow in a fully saturated soil.

7. According to King (1964), the vertical dimension of a horizontal tube of soil (used to check the square root of time law) should be small compared to the value of $p_d/\rho_w g$. Describe some difficulties that would be expected if this condition were not satisfied.

8. Describe an important mathematical advantage of writing a flux equation for an infiltration analysis in terms of the "total" flux q_t.

9. Would the same mathematical advantage apply for an infiltration process involving 2- or 3-dimensional flow? Explain.

10. Explain why the "fractional flow" analysis employed by Morel-Seytoux for infiltration problems is more effective for determining infiltration rates than for determining saturation distributions.

11. Explain why a flux equation in the form of the Green and Ampt flux equation can be exact, whereas an equation combining this flux equation with a continuity equation is usually not exact.

12. When infiltrometers are used to measure infiltration rates in a field overlying a high water table, it is possible that the measured rates may be significantly higher than those observed during a flooding operation. Explain.

13. Describe a set of conditions under which infiltration may be described by the "diffusivity" equation with negligible error.

14. Describe a set of conditions under which it is more appropriate to use a 2-phase flow analysis of the infiltration process.

15. Consider an experiment in which a small square plot of ground, say four square meters, is exposed to a small constant rate input of water (from a sprinkler) slightly less than K_m. Assuming that a water table exists at a depth of 2 meters, would the 2-phase flow equations of McWhorter, accounting for air compression, give a better evaluation of ponding time than the solution of Parlange for the "diffusivity" equation? Explain.

16. Consider a case of drainage from an initially fully saturated column of sand. The water table is suddenly lowered to the base of the column at time $t = 0$. Assuming that the water table is maintained at this depth throughout the period of observation, derive an equation (based on an "upside down" Green and Ampt analysis) for the discharge rate as a function of the cumulative discharge at a particular time.

17. Discuss the possible consequences of analyzing drainage for the case described in problem 16 by assuming the distribution of water above the point where $p_c = p_e$ to be a static distribution. Would you expect this assumption to provide a more or less accurate description of the discharge rate as a function of the cumulative discharge than that obtained from the "upside down" Green and Ampt analysis? Explain.

CHAPTER VI

MODEL SIMILARITY

6.1. MODELS

Only a small fraction of the situations encountered in practice are entirely amenable to analytical solutions. More often, the boundary conditions are too complex to be expressed in simple mathematical terms. In such cases, it is necessary to drastically idealize the problem or to obtain solutions by using a finite difference or finite element model, and perhaps to check the results with an appropriately scaled physical model.

An appropriate model, whether physical or numerical, must create the same relationship among the pertinent variables as does the system being analyzed, i.e., the prototype. In other words, the model must be similar to the prototype. Numerical models may require considerable time to formulate and substantial expense for computer time. Consequently, it is desirable to obtain results in a form that is applicable to as large a range of field conditions as is practical. To accomplish this, the governing equations (and the initial and boundary conditions) should be expressed in terms of such variables that the model is similar to the largest possible range of physical situations. This involves choosing the most appropriate set of variables to describe the system under consideration and scaling the variables with appropriate scale factors.

Physical models usually require an even larger expense for construction and operation than do numerical models. Therefore, it is desirable to determine the least restrictive criteria of similarity so that the model is as small and inexpensive to construct and operate as possible, as well as being similar to as large a range of field situations as is practical.

Therefore, the topics discussed in this Chapter include:

(1) The definition of similarity,
(2) The selection of variables,
(3) The selection of appropriate scale factors,
(4) Finding the least restrictive criteria of similarity,
(5) Application of the similarity principles to flow in porous media.

6.2 DEFINITION OF SIMILARITY

The technical meaning of "similarity" is somewhat more restricted than its non-technical meaning, because it refers to a specified closed set of variables and not to every conceivable property. The variables needed for a description of a fluid system might include pressure, density and velocity but not color, for example. Two systems are similar in a technical sense when the same relationship exists among the specified variables after they have been transformed in an appropriate manner.

The size of the set of systems that are similar depends in part upon what variables have been selected to describe a system, and in part upon how the variables are transformed. An objective of the transformation process is to maximize the number of similar systems. An explanation of how this is done is perhaps more easily understood when the variable under consideration is a geometric variable. In any case, an understanding of geometric similarity is essential before other kinds of similarity can be explained.

6.2.1 Geometric similarity - When investigating the similarity of physical systems, the geometry of boundaries that enclose the systems must be considered. Spaces enclosed by boundaries of the same shape are geometrically similar regardless of their size. For example, every sphere is similar to every other sphere regardless of the length of its radius. Likewise every cube is similar to every other cube regardless of the dimension of its sides.

On the other hand, a prolate spheroid (generated by rotating an ellipse about its major axis) is not necessarily similar to other prolate spheroids. The shape of such a space depends upon the ratio of two characteristic dimensions, i.e., the lengths of the major and minor axes. If the ratio of these two dimensions is equal, the spheroids are similar. Furthermore, if two prolate spheroids are similar, the ratio of lengths of the major axes for these two spheroids will be the same as the ratio of lengths of the minor axes.

The length of one of the axes (say the minor axis) can be selected as a common "scale factor" with which to measure the "scaled length" of any other "corresponding dimension", e.g., the major axis. If prolate spheroids are similar, the scaled lengths of corresponding dimensions are equal.

An analogous relationship can be used to define geometric similarity of spaces with boundaries of any shape, regardless of their conformation or symmetry. However, an infinite set of dimensions is needed to specify the shape in the general case. Furthermore, the meaning of the term "corresponding" is not as obvious as for prolate spheroids. In any case, the word "corresponding" has meaning only in respect to similar geometric spaces regardless of their shape.

It is possible to locate an origin for a coordinate frame of reference within the space enclosed by the boundaries of any system. If the spaces are similar, it will be conceptually possible to identify a common point for an origin of a frame of reference in each of the spaces, say at the centroid. It also will be conceptually possible to orient the coordinate axes so that the distances to the boundary from the origin (in any particular direction with respect to the coordinate frame) will be the same ratio of a common "scale factor". A coordinate frame of reference consistent with this condition is called a corresponding frame of reference. Therefore, similar geometric spaces are those for which corresponding frames of reference exist.

A geometric scale factor is a characteristic dimension of the spaces being compared. For example, a suitable scale factor could be the length of a line segment from the origin of a corresponding frame of reference to the boundary in a particular direction. With such a scale factor, the scaled distance to the boundary from the origin in any given direction is the same for all geometrically similar spaces.

Points located at the same scaled distance in the same direction from the origin of a corresponding frame of reference are called corresponding points. The lengths of line segments connecting corresponding points are called corresponding dimensions. Any corresponding dimension is a valid scale factor, because it is a characteristic length dimension of the spaces being compared.

There is an infinite set of factors consistent with the definition of a geometric scale factor given above. In choosing a particular member from this set, it is appropriate to select one that can be identified without ambiguity and which can be measured easily with adequate precision.

In some particular models, it may not be necessary to maintain similarity of geometry in respect to three coordinate directions. For example, if it is known that a flow system can be characterized adequately as a 1-dimensional vertical flow problem, it is not necessary to construct a model having complete geometric similarity with a prototype. The prototype might consist of soil beneath a flooded rice paddy, but the model could be a small-diameter column of soil, only the depth dimension in relation to soil properties being important. Usually, it is necessary to maintain geometric similarity only for those dimensions that are included in a differential equation describing the interrelation among the variables. This is one way in which an appropriate selection of variables can make the requirements for similarity less restrictive.

When investigating geometric similarity for some systems, similarity of more than one kind of boundary must be considered. For example, when dealing with two fluids in a porous matrix, it is necessary to consider (at least implicitly) the geometry of a particular fluid with boundaries consisting of the internal solid surfaces and fluid interfaces, as well as the external boundary of the porous matrix. The need to consider the internal geometry explicitly on a microscopic scale may sometimes be avoided by selecting "macroscopic" variables to describe the system, where the latter variables are volume averages of corresponding "microscopic" variables. This is another way an appropriate selection of variables can make the requirements for similarity less restrictive.

When the boundaries of a system are changing with time, a more restricted definition of the characteristic dimension used as a scale factor may be necessary. For example, it may be necessary to define the scale factor as the length of a line segment between two corresponding points at a particular scaled time. The question of how time is scaled is discussed in the following section. An example of a system for which the boundaries change with time is unsteady flow in an open channel.

Similarity with respect to a geometric variable in the unsteady case implies that the system is geometrically similar at any particular scaled time. However, the space enclosed by the boundaries at some particular clock time will not be similar unless the systems are identical. For example, the geometries of two similar flow systems in open channels (in which traveling waves exist) will not be similar when compared at a common clock time unless they are identical channels with identical boundaries and identical conditions. Otherwise, the geometries will be similar only at a common scaled time.

6.2.2 Similarity of non-geometric variables - Similarity in respect to a non-geometric variable implies that the scaled values of the variable are equal at corresponding points at corresponding times. This definition applies only for systems that remain geometrically similar at any scaled time, because the definition is in reference to conditions at corresponding points, which exist only in geometrically similar spaces.

Corresponding times are times having equal scaled values. Similarity in respect to time means that equal values of other scaled variables occur at corresponding points at equal scaled times. However, the definitions of similarity given here assume that the scale factor used for each variable is a characteristic parameter of the system as described in the following section.

6.2.3 Characteristic parameters - As explained in Section 6.2.1, a characteristic length parameter is the length of a line segment that connects any two corresponding points in a corresponding frame of reference. Consequently, an infinite set of suitable scale factors exist for length. However, if a particular length scale factor is a characteristic parameter of the system, all other suitable scale factors for length are necessarily directly proportional to that factor in similar systems. For example, the radius of a pipe is a characteristic length dimension for pipes. Consequently, the internal diameter is also a characteristic length dimension for pipes, because it is always twice the radius. However, the outer diameter of the pipe is not a characteristic parameter for the fluid system, and its magnitude is not twice the radius of the flowing column of fluid. The use of the latter quantity for scaling merely changes the unit of measurement and serves no useful purpose.

Other variables may differ from length in that they are assigned directions as well as magnitude. For example, position, velocity and acceleration are quantities which are characterized by direction as well as magnitude. However, only the magnitudes are scaled. Two scaled vectors are equal if their directions as well as their scaled magnitudes are equal.

Variables may be classified also according to whether or not they are assigned values at a particular point. A variable which varies with respect to position in a system is called an intensive variable. The vectors mentioned above are examples of intensive variables. Pressure and temperature are examples of intensive variables which are scalar quantities.

Similarity with respect to an intensive variable implies that the "distribution" of the variable in both space and time is similar. Any non-zero magnitude of the variable at a corresponding point at a corresponding time is a suitable characteristic parameter for such a variable. For example, if the velocity vector during steady laminar flow in a tube with a circular cross section is scaled with the magnitude of the velocity at the centerline, the scaled velocity vector will be the same at equal radii in all such tubes. This is because the direction of flow at all radii is parallel to the axis, and the distribution of the magnitude of the velocity is similar in respect to space in all circular tubes during laminar flow. The average velocity Q/A in the tube can also be used as a velocity scale factor because this quantity is directly proportional to the velocity at the centerline for all tubes with similar velocity distributions. The velocity at any corresponding radius, except at the boundary, could also be used for this purpose. The velocity at the boundary could not be used, because it has a zero magnitude.

6.2.4. Explicit and derived scale factors - If x_o is a characteristic length of a system and u_o is a characteristic velocity, the ratio x_o/u_o is a characteristic time parameter for the same system. Furthermore, any factor having this dimension which is directly proportional to x_o/u_o is also a characteristic time parameter and can be used as a time scale factor. The parameters x_o and u_o are examples of "explicit" scale factors whereas x_o/u_o is an example of a "derived" scale factor. The derived scale factors are combinations of other scale factors and constant system parameters having the appropriate dimensions.

The number of explicit scale factors required is not greater than the number of basic dimensions needed to express all of the variables in a given set. For most fluid flow systems, the number of basic dimensions is not more than three, i.e., force, length and time. Often the number is less than this. Explicit scale factors normally are used to scale those variables for which valid scale factors are clearly indicated. For example, the depth of water in an open channel or the diameter of a pipe is selected to scale lengths, and the maximum permeability of a soil is selected as an explicit scale factor for effective permeabilities. Variables for which no obvious characteristic parameter exists are usually scaled with derived scale factors. The procedure for systematically choosing appropriate derived scale factors is described in the following sections.

6.3 CRITERIA OF SIMILARITY

6.3.1 Selection of variables - Establishing criteria of similarity for any type of system begins with the identification of a set of variables that describe those aspects of the system that are of interest to the investigator. It may happen that the system can be viewed from more than one perspective and consequently alternative sets of variables may be selected to describe the system. Normally, the set is selected for which the criteria of similarity are the least restrictive and for which the criteria can most easily be established.

For example, flow of two fluids in a porous matrix may be considered from either a "microscopic" or a "macroscopic" viewpoint. With the former viewpoint, it is necessary to consider the force and material balances on microscopic reference elements of each individual fluid. Moreover, the boundaries of each fluid consist partly of the internal solid surface and partly of a changing interface between the two fluids.

With a macroscopic viewpoint, the reference elements are macroscopic volume elements of the porous matrix, so that the force and material balances refer to volume averages over the macroscopic elements. The characteristics of the internal geometry are accounted for implicitly, through the functional relationships between fluid content, permeability and pressure difference across interfaces.

Inevitably, the macroscopic viewpoint results in a shorter list of variables and system parameters and a more easily established set of similarity requirements. Describing a particular process in greater detail, or in greater generality, than required usually results in more restrictive criteria of similarity than necessary.

6.3.2 Transformation of differential equations - It is possible to develop criteria of similarity from a given set of variables by dimensional considerations alone [Buckingham, E. (1914)]. The process is called dimensional analysis. However, the criteria can be developed more directly, and usually in a more useful form, by beginning with a differential equation (or equations) relating the variables.

In conformance with the definition of similarity previously stated, similar systems are those for which the scaled governing equations and conditions yield identical particular solutions. It is usually possible to determine (by inspection of the scaled equation) under what conditions identical particular solutions exist.

Equation 3.1, for flow of an incompressible Newtonian fluid, is considered as an illustration of how a differential equation can be used to determine criteria of similarity. This equation is

$$\rho \frac{du_i}{dt} = -\rho g \sin\beta - \frac{\partial p}{\partial x_i} + \mu \frac{\partial^2 u_i}{\partial x_j \partial x_j} ,$$

where β is the angle of the direction i with respect to a horizontal plane. Equation 3.1 is sufficient to describe steady flow of a single Newtonian fluid of constant density and viscosity with fixed boundaries. If unsteady flow and variable boundaries are to be considered, a material balance equation would also be required to describe the system.

Assuming that Equation 3.1, as written, is sufficient to describe the system, the process of determining criteria of similarity is begun by scaling the variables in the equation. The unscaled variables are x_i, u_i, p and t. Note that the appearance of time as a variable in this set does not imply that the system is necessarily unsteady. Nevertheless, time is a variable that needs to be scaled. The scale factors for the variables are tentatively identified only as x_o, u_o, p_o and t_o.

The scaling process involves transforming the unscaled equation, not changing it. Therefore, when a variable is divided by a particular scale factor, the term in which it occurs is multiplied by the same factor to the power to which the variable appears in the term. The result of the scaling process is

$$\frac{d\hat{u}}{d\hat{t}} = -\left(\frac{t_o g}{u_o}\right) \sin\beta - \left(\frac{t_o p_o}{x_o u_o \rho}\right) \frac{\partial \hat{p}}{\partial \hat{x}_i} + \left(\frac{\mu t_o u_o}{u_o \rho x_o^2}\right) \frac{\partial^2 \hat{u}_i}{\partial \hat{x}_j \partial \hat{x}_j} ,$$

where the scaled variables are designated by the symbol "^" over the variable. The factor t_o can be eliminated as an explicit scale factor by defining $t_o \equiv x_o/u_o$. This is desirable because an obvious explicit scale factor for time does not exist. The result is

$$\frac{d\hat{u}}{d\hat{t}} = -\left(\frac{x_o g}{u_o^2}\right) \sin\beta - \left(\frac{p_o}{\rho u_o^2}\right) \frac{\partial \hat{p}}{\partial \hat{x}_i} + \left(\frac{\mu}{\rho x_o u_o}\right) \frac{\partial^2 \hat{u}_i}{\partial \hat{x}_j \partial \hat{x}_j} .$$

Further simplifications of the scaled equation cannot be made in this case. Consequently, explicit scale factors for length, velocity and pressure are required.

The scaled version of Equation 3.1 is identical to the unscaled version regardless of how the factors x_o, u_o and p_o are selected. However, no advantage is obtained from writing it in the scaled form unless the factors used are characteristic parameters of the system as described in Sections 6.2.1 and 6.2.3.

To understand the advantage gained by properly scaling Equation 3.1, the requirements for similarity in terms of the unscaled variables x_i, u_i, p and t are considered first. An examination of the unscaled equation shows that the requirements are:

(1) The fluid properties ρ and μ must be identical.
(2) The angles β must be equal.
(3) The boundary and initial conditions in terms of the unscaled variables must be the same, and this implies that the systems must be the same size.

An examination of the scaled version of Equation 3.1 shows that the requirements for similarity in terms of the scaled variables are:

(1) The three constant dimensionless coefficients appearing in the scaled equation must be equal.

(2) The boundary and initial conditions in terms of scaled variables must be the same. However, this does not imply that the size of the systems are necessarily equal.

An examination of the two sets of similarity criteria shows that an important advantage of scaling the variables is that similarity can be obtained without identical sizes and identical fluids. This is important because it is rarely practical to build a physical model with the identical size of a prototype system. Furthermore, the time required to observe corresponding behavior can be greatly reduced, sometimes by several orders of magnitude, by reducing the model size and adjusting other variables so that the ratio x_o/u_o is reduced.

When the scale factors x_o, u_o and p_o are selected from characteristic parameters (as described in Section 6.2.3) the coefficients in the scaled version of Equation 3.1 can be associated with well-known criteria of similarity as follows:

(1) $$\left(\frac{gx_o}{u_o^2}\right)^{-1/2} = \frac{u_o}{\sqrt{x_o g}} \equiv \text{Froude number}, \qquad 6.1$$

(2) $$\left(\frac{p_o}{\rho u_o^2}\right)^{-1/2} = \frac{u_o}{\sqrt{p_o/\rho}} \equiv \text{Euler number}, \qquad 6.2$$

(3) $$\left(\frac{\mu}{\rho u_o x_o}\right)^{-1} = \frac{\rho u_o x_o}{\mu} \equiv \text{Reynolds number}. \qquad 6.3$$

The coefficients can be transformed by being taken to any power (positive, negative or fractional) if the purpose is only to serve as criteria of similarity, since the equality of such numbers ensures the equality of the coefficients appearing in the scaled form of Equation 3.1. However, to be properly designated by the names indicated in Equations 6.1, 6.2 and 6.3, they should have the form shown.

6.3.3 Similarity for special cases - Criteria of similarity for a particular type of flow system can be determined by starting with an equation that describes the flow with as few simplifying assumptions as possible. However, this procedure may lead to the inclusion of a number of similarity criteria that have little or negligible significance in a particular case.

A more satisfactory procedure is to start with an equation which is specific for the type of system under consideration and which involves all of the simplifying assumptions that have been found to be justified. Scaling the specific equation leads directly to the criteria of similarity which are essential for a particular case.

To illustrate the latter point, a case of steady flow in an open channel with a varying slope is considered. Equation 3.1 for flow of an incompressible Newtonian viscous fluid applies to this type of flow. This equation is derived by balancing forces on a reference element consisting of a microscopic volume of fluid within the flowing stream. Scaling this equation leads to the conclusion that the equality of the Froude, Euler, and Reynolds numbers is each a criterion of similarity, as explained in Section 6.3.2. Experienced investigators of open channel flow recognize that equality of the Froude number is the key criterion in this case whereas the Euler and Reynolds numbers usually are of minor significance. However, there is no "a priori" reason for an investigator to discard the Euler and Reynolds numbers while retaining the Froude number, based on this theory alone.

A specific equation for open channel flow is derived by balancing forces on a macroscopic volume of water that is enclosed in a short section of the channel. A key assumption is made that the flow can be described adequately in terms of an average velocity at a cross section, i.e., Q/A. This is a reasonable assumption because the flow is usually fully turbulent in open channels. The resulting equation is

$$\frac{\partial y}{\partial x} + \frac{v}{g}\frac{\partial v}{\partial x} + \frac{1}{g}\frac{\partial v}{\partial t} = S - S_f \quad ,$$

in which x is a coordinate in the direction of flow, y is the depth of flow above the base of the channel, v is the average velocity, S is the slope of the bed and S_f is the "friction slope". Scaling this equation results in

$$\frac{\partial \hat{y}}{\partial \hat{x}} + \left(\frac{v_o^2}{x_o g}\right) \hat{v}\, \frac{\partial \hat{v}}{\partial \hat{x}} + \left(\frac{v_o^2}{x_o g}\right) \frac{\partial \hat{v}}{\partial \hat{t}} = S - S_f \quad ,$$

where the scale factors x_o, v_o and t_o are selected as explained in Sections 6.2.1 and 6.2.2. The scaled version of this equation shows clearly that the Froude number is the significant criterion of similarity for steady flow in open channels.

An analogous comparison can be made for identifying criteria of similarity in the case of 2-phase fluid systems in porous media. It is usually preferrable to start with the most specific equation that will adequately describe the system being investigated. The significant similarity criteria are more easily deduced from the scaled version of the specific equation.

6.4 SIMILARITY FOR FLOW OF TWO FLUIDS IN POROUS MEDIA

Since flow in porous media involves Newtonian viscous fluids undergoing negligible divergence, the criteria of similarity deduced from Equation 3.1 theoretically should be applicable in this case. In applying these criteria, however, several complexities of 2-phase flow systems must be considered. One complication is that the Navier-Stokes equation must be applied separately to each of the fluid phases so that the criteria of similarity deduced from it for each phase must be combined.

Furthermore, the boundaries of the flow system pertinent to solutions of the Navier-Stokes equation are the internal solid surfaces of the pore space and the interfaces between the two fluid phases. The microscopic boundaries, therefore, vary with saturation. The saturation varies with the capillary pressure and is influenced by surface tension and by wettability as well as by pore geometry. This situation requires the introduction of additional equations and additional criteria of similarity.

The microscopic approach results in criteria which are very difficult to satisfy and may, in fact, be mutually incompatible. However, many of the early investigators of similarity for porous media included microscopic variables in their analyses [Leverett, M. D. et al. (1942)]. Furthermore, most of the early investigators deduced their criteria of similarity by dimensional analysis alone.

The Navier-Stokes equation can be simplified by dropping the acceleration term. This is a reasonable simplification since fluid inertia is rarely a significant factor in 2-phase flow in porous media. Dropping the acceleration term eliminates Reynolds number from the list of similarity criteria. A further simplification can be made by noting that ρ occurs in the remaining terms only as the product ρg so that the two variables can be treated as one, as was pointed out by Loomis and Crowell (1964).

Miller and Miller (1956) were the first to investigate similarity specifically for air-water systems in soils. They approached the problem mainly from a microscopic perspective, although Darcy's equation was considered in their analysis. Their approach leads to a concept of "similar media" which requires the internal pore geometry to be similar, i.e., to differ only by a constant scale factor. When media are similar in this sense they also have equal porosities.

Brooks and Corey (1966) presented a set of criteria without explicit reference to any microscopic variables. Their approach is entirely macroscopic, proceeding directly from the Richards equation. This theory was later extended by McWhorter and Corey (1967) to include systems for which flow of the nonwetting phase cannot be ignored. The latter development is based upon the 2-phase flow equation

$$-\phi \frac{\partial S}{\partial t} = \frac{\partial}{\partial x_i} \left[\frac{k_w}{\mu_w} \left(\frac{\partial p_c}{\partial x_i} - \Delta\rho g \frac{\partial h}{\partial x_i}\right)\right] + \frac{\partial}{\partial x_i} \left[q_{nwi} \left(\frac{\mu_{nw}}{\mu_w} \frac{k_w}{k_{nw}}\right)\right] \quad .$$

The variables in this equation are S, k_w, k_{nw}, x_i, h, p_c, q_{nwi} and t. The saturation S is a wetting phase content θ which has been divided by ϕ, so that $0 \leqq S \leqq 1$. The permeabilities k_w and k_{nw} are scaled by k because this is a maximum value for both k_w and k_{nw}. Consequently, both k_w and k_{nw} vary from 0 to 1.0. The scale factor for p_c is taken as some identifiable value of p_c on the curve of $S = f(p_c)$. McWhorter and Corey (1967) used p_d for this factor, but p_e is equally satisfactory.

The same scale factor for length x_o is used for x_i and h, since h is the vertical coordinate. This factor along with the scale factors for flux rate q_o and time t_o are not identified until after the equation is scaled. The factors are then selected so that as many of the constant coefficients as possible are set equal to 1.0. In this case, all of the constant coefficients, except the viscosity ratio, can be set equal to 1.0 by defining

$$x_o \equiv p_d/\Delta\rho g \quad ,$$

$$q_o \equiv k\Delta\rho g/\mu_w \quad ,$$

and

$$t_o \equiv p_d \mu_w \phi/k(\Delta\rho g)^2 \quad .$$

The scaled version of the 2-phase flow equation that results from this process is

$$-\frac{\partial S}{\partial \hat{t}} = \frac{\partial}{\partial \hat{x}_i} \left[\hat{k}_w \left(\frac{\partial \hat{p}_c}{\partial \hat{x}_i} - \frac{\partial \hat{h}}{\partial \hat{x}_i}\right)\right] + \frac{\mu_{nw}}{\mu_w} \frac{\partial}{\partial \hat{x}_i} \left(\hat{q}_{nwi} \frac{\hat{k}_w}{\hat{k}_{nw}}\right) \quad .$$

The similarity requirements for the general case of 2-phase flow deduced from this equation are:

(1) The initial and boundary conditions in terms of scaled variables are the same. This implies that the external boundaries of the porous matrix are the same shape, and that the sizes in terms of scaled dimensions are equal.

(2) The functional relationships among S, $\hat{p}_c$, $\hat{k}_w$ and $\hat{k}_{nw}$ are identical.

(3) The orientation of corresponding dimensions with respect to the Earth's gravitational field is the same, because the term $\partial\hat{h}/\partial\hat{x}$ appears in the scaled equation.

(4) The viscosity ratio for the two flowing fluids is identical.

6.4.1 Comparison of scale factors for 2-phase flow systems - A majority of the investigators, including both soil and petroleum scientists, who have studied similarity requirements for 2-phase flow in porous media have presented a list of criteria equivalent to that of McWhorter and Corey (1967). However, the soil scientists usually do not consider equality of the viscosity ratio as a requirement because they believe this ratio to be negligible for air-water systems. For the same reason, $\hat{k}_{nw}$ is not one of the variables considered by soil scientists.

The earlier investigators of similarity in the petroleum industry often included a criterion evaluating inertial effects. Later investigators [Rapoport (1955), and Loomis and Crowell (1964)] in the petroleum field pointed out that this is usually not necessary. Miller and Miller (1956) included a requirement that the porosities of the media be equal, because the volumetric water content θ, rather than the saturation S, was used for characterizing fluid concentrations.

Although authors usually agree in respect to explicit statements of similarity requirements, their criteria are in reference to different sets of scaled variables. Therefore, an examination of the scaling factors employed by various authors is needed for an adequate comparison of the similarity criteria that have been proposed. Furthermore, the usefulness of a particular scheme depends upon how easily and precisely the scale factors can be identified and evaluated, as well as upon how precisely the relationship among variables coalesce. Table 6.1 presents a comparison of scale factors used by two petroleum scientists with those proposed by Brooks and Corey (1966) and Miller and Miller (1956). The symbols used in the original papers have been replaced (where possible) by corresponding symbols used elsewhere in this text. The symbol L

Table 6.1. Scale factors of several investigators [McWhorter (1966)].

Unit	Rapoport (1955)	Richardson (1961)	Miller and Miller (1956)	Brooks and Corey (1966)
Length	$\frac{\sigma}{\Delta\gamma}/\sqrt{k/\phi}$	$\frac{\sigma f(\alpha)}{\Delta\gamma}/\sqrt{k/\phi}$	$\frac{\sigma}{\Delta\gamma d}$	$\frac{p_d}{\Delta\gamma}$
Capillary pressure	$\sigma/\sqrt{k/\phi}$	$\sigma f(\alpha)/\sqrt{k/\phi}$	σ/d	p_d
Time	$\frac{\mu L\phi}{k\Delta\gamma}$	$\frac{\mu L\phi}{k\Delta\gamma}$	$\frac{\mu L}{\Delta\gamma\ d^2}$	$\frac{\mu p_d \phi_e}{K(\Delta\gamma)^2}$
Flux rate	$\frac{k\Delta\gamma}{\mu}$	$\frac{k\Delta\gamma}{\mu}$	$\frac{\Delta\gamma d^2}{\mu}$	$\frac{k\Delta\gamma}{\mu}$

refers to a macroscopic length which characterizes the size of the entire flow system, $f(\alpha)$ is a function of the contact angle of the interfaces and $\Delta\gamma$ designates $\Delta\rho g$.

It is noteworthy that the scale factors shown in Table 6-1 for macroscopic length are, in every case, equal to the scale factor for capillary pressure divided by the difference in specific weights of the two fluids involved. Furthermore, the scale factors are inversely proportional to a characteristic pore dimension. The characteristic pore dimension is represented by $\Delta\rho g/p_d$ in the scale factor of Brooks and Corey (1966) and by a function of $\sqrt{k/\phi}$ in the petroleum literature. Only Miller and Miller (1956) have included the characteristic pore dimension explicitly.

According to the theory presented in Section 2.4 leading to Equation 2.12, if d is assumed to be proportional to a maximum value of A/wp, it is also proportional to $\sigma \cos \alpha/p_d$. If α is a constant, this implies that σ/d is proportional to p_d. Furthermore, according to the theory presented in Section 3.6.1, p_d is inversely proportional to $\sqrt{k/\phi}$, provided the parameters σ, α, T, k_s and S_r are approximately equal. To the extent that this theory is valid, all of the scale factors presented in Table 6-1 for both capillary pressure and length are equally valid because they are mutually proportional. However, p_d should be an appropriate scale factor for

p_c regardless of the validity of the theory involving the microscopic variables.

It is informative to consider why the scale factor for macroscopic length should be inversely proportional to a characteristic pore dimension. The reason becomes clear if one considers a simple 1-dimensional static case where two fluids (say air and water) exist as continuous phases in a column of soil with a water table ($p_c = 0$) at the bottom of the column. The soil column will be nearly fully saturated with water above the water table for a height that depends upon the magnitude of $p_d/\Delta\rho g$. Clearly, if two such columns are to have similar distributions of water content along their length, the ratio of the overall length L of the column to $p_d/\Delta\rho g$ must be the same. Note that p_d is inversely related to the size of the pores. However, as pointed out above, the exact relationship depends upon some variables that are not easily measured as they exist within the pore space, e.g., σ, α and A/wp. Consequently, $p_d/\Delta\rho g$ seems to be the more appropriate scale factor for length since it is a measurement of the factor directly affecting similarity of behavior.

Leverett et al. (1942) were the first to employ the parameter $\sigma/\sqrt{k/\phi}$ as a scale factor for p_c. This results in

$$J(s) \equiv \frac{p_c}{\sigma} \sqrt{k/\phi}$$

being one of the variables often used to describe flow in petroleum reservoirs. The variable J(s) is often called the "J" function in petroleum literature.

The only difference between the scale factors (of the two petroleum scientists) shown in Table 6-1 is that Richardson (1961) has introduced a factor accounting for the effect of the contact angle. Perkins and Collins (1960), whose scale factors are not shown in Table 6-1, proposed essentially the same set of scale factors except that they replaced k with k_m, the permeability with entrapped nonwetting fluid.

The scale factors for time, with the exception of that of Miller and Miller (1956), are essentially the same. The time scale factor of Brooks and Corey (1966) includes $p_d/\Delta\rho g$ rather than L but this is not a significant difference, since their theory requires L to be proportional to $p_d/\Delta\rho g$ for similar systems. Miller and Miller (1956) are the only authors in this group who have not included ϕ in the time scale factor, since they have used the volumetric water content θ, rather than the saturation S, as a measure of wetting phase concentration. Earlier investigators in the petroleum industry [Leverett et al. (1947)] also regarded the equality of ϕ to be a criterion for similarity, but this is not usually mentioned as a requirement by later authors in the petroleum field.

The scale factors for flux rate shown in Table 6-1 also are identical, except for that of Miller and Miller (1956) who have introduced the parameter d in this scale factor rather than the permeability k. The validity of the use of d in the scale factors for both flux rate and time depends upon the validity of the theory presented in Section 3.6.1. The scale factors for flux rate and time in terms of k and ϕ should be appropriate in any case.

A question arises as to whether ϕ or ϕ_e should appear in the time scale factor. It would appear from theoretical considerations that where some portion of the pore space does not contribute to storage or the flow process, it is appropriate to use ϕ_e rather than ϕ. However, there are situations, e.g., infiltration into air dry soil, where part of the pore space normally occupied by residual water does contribute to storage, although it may not contribute significantly to the flow capacity. In the latter case, it is possible that the equality of S_r may become an additional requirement of similarity, and it may be more appropriate to use ϕ in the time scale factor rather than ϕ_e.

6.4.2 Indentification of similar media - The criterion of similarity most difficult to satisfy in particular cases is the requirement that the functional relationship among S, $\hat{p}_c$, $\hat{k}_w$ and $\hat{k}_{nw}$ be the same. According to the Brooks and Corey (1966) similarity theory, when this requirement is satisfied the porous media are considered similar. This is in contrast with the theory of Miller and Miller (1956) who regarded media to be similar only when their internal pore space was geometrically similar and when the pore space occupied an equal fraction of the entire matrix volume.

By either of the above definitions of similar media, the requirements for similarity in the most general case is very difficult to satisfy. This is a problem that was investigated extensively by Brooks and Corey (1966). They found that similarity, according to their definition, can be satisfied without too much difficulty in some special cases. These cases include those for which the media are homogeneous as well as isotropic; saturation hysteresis does not occur, and the saturation of the wetting fluid is never reduced below the residual saturation.

It was found that for such cases, the functional relationships among $\hat{k}_w$, $\hat{k}_{nw}$, S_e and $\hat{p}_c$ are the same for all media having similar pore-size distributions, as indicated by the equality of the index λ, where λ is the empirical exponent relating S_e to p_c (see Section 2.3.6).

It should be noted that the replacement of S by S_e in the Brooks-Corey theory eliminates the equality of S_r as a necessary criterion of similarity. However, it also reduces the applicability of the theory to systems in which S is everywhere greater than S_r.

According to the theory of Brooks and Corey (1966) it should be possible to identify similar porous media simply by determining curves of $S = f(p_c)$ on undisturbed samples of the media. This is because $\hat{k}_w$, $\hat{k}_{nw}$, $\hat{p}_c$ and $\hat{S}_e$ are all interrelated and all depend upon the pore-size distribution.

The Brooks and Corey (1966) theory of similarity has been extended to apply to stratified sands by G. L. Corey (1965) and Taleghani (1972). However, the problem of identifying similarity requirements for media undergoing hysteresis and for non-isotropic media are problems needing much more research.

6.4.3 Similarity for special cases of flow in porous media - It was pointed out in Section 6.3.3 that usually it is easier to identify the significant criteria of similarity for special cases from equations that are derived specifically for such cases. This generalization applies to 2-phase fluid systems in porous media as well as to other problems of fluid flow.

An example is the case of linear imbibition of a liquid into a horizontal tube of homogeneous soil at some initial liquid content (see Section 5.2). The source of the liquid is at one end of the tube and the pressure of the liquid at the source is equal to or slightly greater than atmospheric pressure. The other end of the tube is open to the atmosphere so that resistance to the escape of air ahead of the advancing liquid is expected to be negligible.

In choosing a specific equation to describe this case, it is important to note that the source pressure is such that a portion of the tube of soil will be fully saturated after a finite interval of time following the start of the imbibition process. Equation 5.5 does not apply in the fully saturated region since θ is not a single-valued function of ψ. The appropriate equation is a simplified version of Equation 5.4; i.e.,

$$\phi \frac{\partial S}{\partial t} = - \frac{\partial}{\partial x} (K \frac{\partial \psi}{\partial x}) \quad .$$

The volumetric water content θ has been replaced by ϕS because this avoids the necessity of retaining the equality of ϕ as a criterion of similarity. The effective porosity ϕ_e is not used because θ_i is not necessarily equal to or greater than θ_r. An alternative procedure is to replace ϕ with an initial porosity ϕ_i given by $\theta_m - \theta_i$, where θ_m and θ_i are the maximum and minimum liquid contents respectively. In the latter case, S is defined as θ/ϕ_i rather than θ/ϕ.

The variables in the unscaled equation are x, K, Ψ, t and S. The scale factor for K is K_m, the maximum value of K, and the scale factor for time t_o is not identified until after the scaling process has been carried out. The resulting equation is

$$\frac{\partial S}{\partial \hat{t}} = - (\frac{t_o K_m}{\phi \psi_o}) \frac{\partial}{\partial \hat{x}} (\hat{K} \frac{\partial \hat{\psi}}{\partial \hat{x}}) \quad .$$

By designating t_o as $\phi\psi_o/K_m$, the coefficient in this equation can be set equal to 1.0 without loss of generality. The resulting scaled specific equation is

$$\frac{\partial S}{\partial \hat{t}} = - \frac{\partial}{\partial \hat{x}} (\hat{K} \frac{\partial \hat{\psi}}{\partial \hat{x}}) \quad .$$

From the scaled specific equation, the following similarity requirements are deduced:

(1) The functional relationships among S, $\hat{\psi}$ and $\hat{K}$ are identical.
(2) The tube lengths are equal ratios of the value of ψ_o for each particular soil; i.e., $\hat{L}$ is equal.
(3) The value of $\hat{\psi}$ at the inflow ends is the same at corresponding times.
(4) The initial value of S is equal in each similar soil tube.

It is not required that the same liquid be imbibed in each soil tube, provided that liquid-solid reactions do not prevent requirement 1 from being satisfied. Moreover, the theory does not specify any conditions on the dimensions of the tube cross-sections, because the imbibition is assumed to be linear in a horizontal direction only. However, it would appear from the observations of King (1964) that the vertical dimension of the tubes must be small compared to $p_d/\rho g$ to ensure that gravity does not affect the fluid distribution (see Section 5.2.1).

The first similarity requirement can be satisfied in this case only if S_r is equal for each soil. However, if it is specified that the initial saturation is equal to or greater than S_r, the first requirement is much less restrictive. In the latter case, ϕ and S can be replaced by ϕ_e and S_e respectively, and the first requirement is satisfied by equal values of λ, i.e., by equal values of the slope of a curve of $\ell n\ S_e = f(\ell n \psi)$ obtained for the wetting cycle starting from S_i.

It is quite likely that an experienced investigator would predict the similarity criteria for this example by referring to the requirements deduced from the general 2-phase flow equation and eliminating those that clearly are not significant. However, in other cases, it may not be obvious what criteria can be safely eliminated. An example of such a case is discussed in the following:

A case is considered in which oil is displaced linearly in a thin stratum of porous rock by an invading front of brine. The brine is injected from a line of wells at a constant rate and fluids are removed from another line of wells at a distance L from the injection wells. It is assumed that the initial saturation of brine in the rock formation is uniform and is equal to or greater than the residual saturation. The displacement takes place under a very large pressure gradient and the slope of the formation is very small.

Equation 5.53 is an appropriate equation for this case, but as explained in Section 5.3.1, it can be written more specifically for the present purpose as

$$\phi_e \frac{\partial S_e}{\partial t} = - q_t \frac{\partial f_w}{\partial x} ,$$

which is a version of the Buckley-Leverett equation. In this case, the total flux rate q_t is also the constant injection rate. Consequently, it can be treated as a parameter of the system which does not vary in time or space.

The unscaled variables are S_e, x, f_w and t. The scale factor for x is necessarily proportional to the size of the system being compared. Consequently, the scale factor for x

is taken as L, the distance between the injection and the producing wells. The scale factor for time is selected so that the constant coefficient is equal to 1.0. The result is

$$\frac{\partial S_e}{\partial \hat{t}} = - \frac{\partial f_w}{\partial \hat{x}}$$

where $\hat{t}$ is $tq_t/\phi_e L$ and $\hat{x}$ is x/L.

The criteria of similarity are:

(1) The relationship $f_w = f(S_e)$ is the same.

(2) The initial effective saturation is the same.

According to the McWhorter and Corey (1967) similarity theory, the first requirement can be satisfied if the porous rocks have equal values of λ, and if the viscosity ratios are equal. It is not possible to specify less restrictive criteria of similarity with the set of variables selected to describe the system. However, by selecting a slightly different set of variables, it is possible to eliminate equality of viscosity ratio as a separate criteria of similarity.

As explained in Section 5.3.1, Equation 5.53 can also be written as

$$\phi_e \frac{\partial S_e}{\partial t} = - \frac{\mu_{nw}}{\mu_w} \frac{\partial}{\partial x} (q_{nw} \frac{k_w}{k_{nw}}) \quad ,$$

where the subscripts "nw" and "w" refer to oil and brine respectively. The new set of unscaled variables are S_e, x, q_{nw}, k_w/k_{nw} and t. In this set, the variable f_w has been replaced by q_{nw} and k_w/k_{nw}. The replacement of one variable by two others may seem, at first, to be a step in the wrong direction. However, this procedure separates the viscosity ratio from the permeability ratio and makes possible the inclusion of the viscosity ratio in the time scale factor.

The scaled version of the specific equation is

$$\frac{\partial S_e}{\partial \hat{t}} = - \frac{\partial}{\partial \hat{x}} (\hat{q}_{nw} \frac{k_w}{k_{nw}}) \quad ,$$

where $\hat{q}_{nw} \equiv q_{nw}/q_w$ and $\hat{t} \equiv \mu_{nw} q_t t/\mu_w L\phi_e$.

The similarity requirements are:

(1) The functional relationship $k_w/k_{nw} = f(S_e)$ is the same.
(2) The initial effective saturation is the same.

The first requirement is satisfied if the porous rocks have equal values of λ. Equality of viscosity ratio is not required.

Evidently, a number of specific equations involving different sets of variables may have to be tried to find the least restrictive set of similarity requirements. Furthermore, the least restrictive set may not always be apparent from an inspection of the set applicable to the most general case of 2-phase flow. For example, it is not likely that an investigator would guess that a sufficient set of similarity criteria for the last example presented would not necessarily include equality of the viscosity ratio, since the viscosity ratio obviously affects this type of displacement process.

6.4.4 Tests of similarity criteria - Several experiments have been made by soil scientists to test the similarity theory of Miller and Miller (1956). These include experiments by Klute and Wilkinson (1958), Elrick et al. (1959) and Wilkinson and Klute (1959). The tests compared results (in terms of transformed variables) of flow experiments with two or more media having contrasting permeabilities and textures but satisfying the criteria for "similar" media. In one case, Elrick et al. (1959) ran hysteresis loops of $p_c(S)$ functions using two different fluids (water and butyl alcohol) on the same media. The agreement was good when an inert medium was used but not as good with a medium containing clay. The varying reaction of the clay to the two fluids undoubtedly prevented the retention functions from behaving in a similar way with each of the fluids. In other cases, however, tests have shown the theory of Miller and Miller (1956) to be valid.

Corey and Corey (1967) reported tests of the Brooks-Corey similarity theory for linear drainage of initially fully saturated soil columns with a zero p_c maintained at the base of the columns. Outflow as a function of time was recorded in terms of transformed as well as standard variables. To obtain maximum precision and reproducibility of measured parameters, Soltrol was used instead of water as the wetting fluid. The lengths of the columns were adjusted to be the same multiple of $p_d/\rho g$ for each of the two materials having the same value of λ. The results are shown in Figures 6-1 and 6-2.

Hedstrom et al. (1971) conducted 2-dimensional drainage experiments in two soil flumes, the larger flume being 12.2 meters long and 1.22 meters high and the smaller flume having

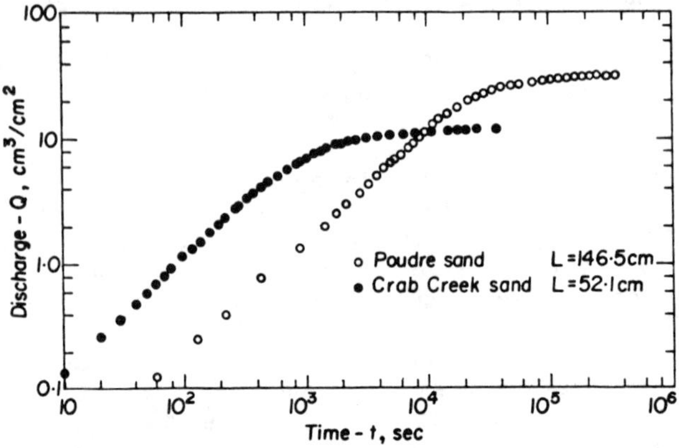

Figure 6-1. Outflow as a function of time (in terms of standard units) from columns of equal scaled height containing media of identical λ. [After Corey and Corey (1967)].

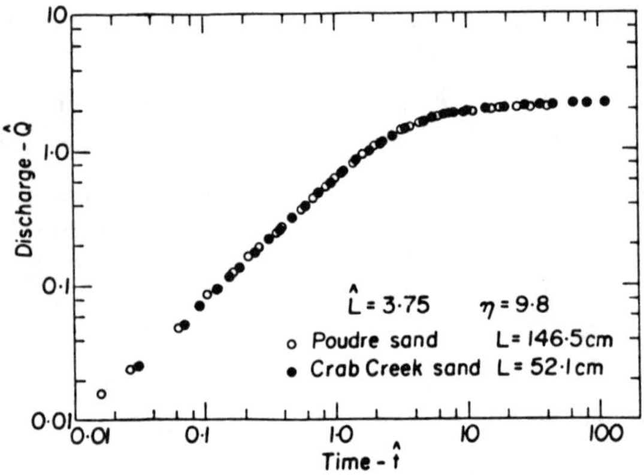

Figure 6-2. Outflow as a function of time (in terms of transformed variables) from columns of equal scaled height containing media of identical λ. [After Corey and Corey (1967)].

dimensions about 1/3 as large. Soltrol was used rather than water in this experiment to improve the stability of the medium structure and to permit maximum precision in measurement of the medium parameters.

Examples of the data obtained for discharge as a function of time in terms of standard units and transformed variables are shown in Figures 6-3 and 6-4 respectively. Data obtained for the elevation of the water table as a function of time also coalesce very closely. From such experiments, Hedstrom et al. (1971) concluded that the similarity criterion is probably valid for the 2-dimensional drainage process, and that results

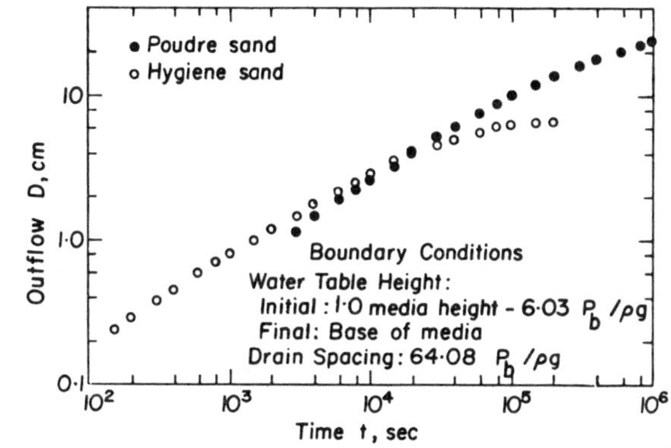

Figure 6-3. Cumulative outflow from 2-dimensional drainage models with similar media. [Hedstrom et al. (1971)].

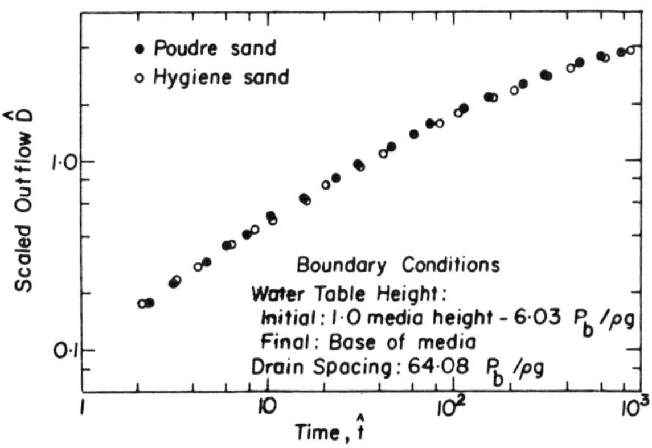

Figure 6-4. Cumulative outflow from 2-dimensional drainage models with similar media (in terms of transformed variables). [Hedstrom et al. (1971)].

from the small flume could be extrapolated to apply to the larger system.

The experiments of Corey and Corey (1967), and Hedstrom et al. (1971) represent cases in which the resistance to flow of a non-wetting phase is undoubtedly negligible. McWhorter and Corey (1967) reported results of a test for a case in which the resistance to flow of the non-wetting phase is significant. Details of this experiment have been presented in a thesis by McWhorter (1966), a brief description of which is given below.

Three columns of porous media were first fully saturated with Soltrol and then drained under the influence of gravity for a period corresponding to a scaled time of 1.0. At this time, a scaled air pressure of 1.2 was applied at the top of each column. The cumulative outflow from each column was measured as a function of time for the entire period. One of the columns was a core of Berea sandstone, 114 cm in length. The other two were shorter columns of unconsolidated sand, all of them, however, having the same height relative to respective values of $p_d/\rho g$. The sandstone core was 3.6 times as long as the shorter of the two unconsolidated sand columns (in terms of standard units) and required 108 times as long to drain to an equivalent saturation. The λ values for the three columns were not greatly different, being 2.8 for two of the columns and 2.1 for the third.

The results of the experiment are presented in Figures 6-5 and 6-6. Note that a sharp increase in liquid discharge occurred at a scaled time of 1.0, as would be expected, because this is when the air pressure was applied. Although the data in terms of standard units are very different for the three columns, they coalesce closely when expressed in terms of transformed variables. Evidently, the index λ provides an adequate criterion of similarity for systems in which a pressure gradient exists in the air phase, at least, for fluids on the drainage cycle.

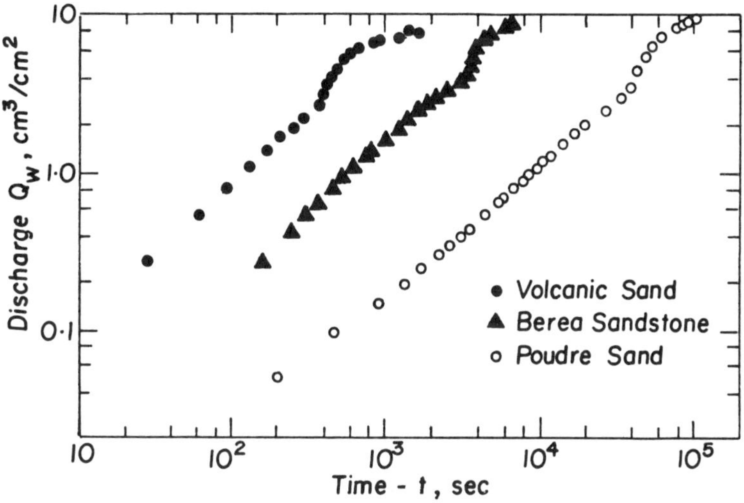

Figure 6-5. Discharge as a function of time for the wetting phase driven by air pressure for three columns of porous media with nearly equal values of λ.

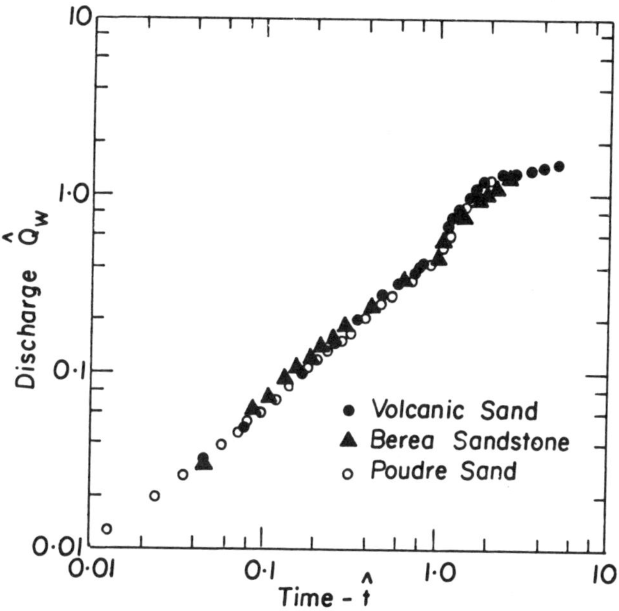

Figure 6-6. Discharge as a function of time (in terms of transformed variables) for the wetting phase driven by air pressure for three columns of porous media with nearly equal values of λ.

REFERENCES

Brooks, R. H. and Corey, A. T. (1966). Properties of porous media affecting fluid flow. J. Irrig. and Drain. Div., Proc. ASCE, IR2. Vol. 92, pp. 61-88.

Buckingham, E. (1914). On physically similar systems. Illustrations of the use of dimensional equations. Physics Review, Vol. IV, No. 4, p. 345.

Corey, G. L. (1965). Similitude for non-steady drainage of partially saturated soils. Ph.D. dissertation, Colorado State University, Fort Collins, Colorado, August, 124 p.

Corey, G. L. and Corey, A. T. (1967). Similitude for drainage of soils. J. Irrig. and Drain. Div., Proc. ASCE, IR3, Vol. 93, pp. 3-23.

Elrick, D. E., Scandrett, J. H., and Miller, E. E. (1959). Tests of capillary flow scaling. Proceedings, SSSA, Vol. 23, No. 5, Sept.-Oct., pp. 329-332.

Hedstrom, W. E., Corey, A. T., and Duke, H. R. (1971). Models for subsurface drainage. Colorado State University, Hydrology Paper No. 48, April, 56 p.

King L. G. (1964). Imbibition of fluids by porous solids. A Ph.D. dissertation, Colorado State University, Fort Collins, Colorado, September, 231 p.

Klute, A. and Wilkinson, G. E. (1958). Some tests of the similar media concept of capillary flow: I, Reduced capillary conductivity and moisture characteristics data. Proceedings, SSSA, Vol. 22, July-Aug., pp. 278-281.

Leverett, M. D., Lewis, W. B. and True, M. E. (1942). Dimensional-model studies of oil-field behavior. Petroleum Technology, T. P. 1413, January, pp. 175-193.

Loomis, A. G., and Crowell, D. C. (1964). Theory and application of dimensional and inspectional analysis to model study displacements in petroleum reservoirs. W. S. Bureau of Mines, Report of Investigations, 6546, 37 p.

McWhorter, D. B. (1966). Similitude for flow of two fluids in porous media. M.S. Thesis, Colorado State University, Fort Collins, Colorado, December, 91 p.

McWhorter, D. B. and Corey, A. T. (1967). Similitude for flow of two fluids in porous media. International Hydrology Symposium, Fort Collins, Sept. 1967, IAHR, pp. 136-140.

Miller, E. E. and Miller, R. D. (1956). Physical theory for capillary flow phenomena. J. of Applied Physics, Vol. 27, No. 4, April, pp. 324-332.

Perkins, F. M. and Collins, R. E. (1960). Scaling laws for laboratory flow models of oil reservoirs. J. of Petroleum Technology, AIME, Technical Note 2063, August, pp. 69-71.

Rapoport, R. A. (1955). Scaling laws for use in design and operation of water-oil flow models. Petroleum Trans., AIME, Vol. 204, pp. 143-150.

Richardson, J. G. (1961). Flow through porous media. Section 16, Handbook of Fluid Dynamics, edited by V. I. Streeter, McGraw-Hill Book Company, Inc., New York, pp. 16-3 through 16-112.

Taleghani, M. S. (1972). Criteria for modeling one-dimensional unsteady drainage of layered soils. M.S. Thesis, University of Manitoba, Winnipeg, Canada, April, 102 p.

Wilkinson, G. E. and Klute, A. (1959). Some tests of the similar media concept of capillary flow: II. Flow Systems Data. Proceedings, SSSA, Vol. 23, No. 6, Nov.-Dec., pp. 434-437.

PROBLEMS AND STUDY QUESTIONS

1. For model studies involving flow in open channels, the Froude number is an important criterion of similarity. This number includes a characteristic length dimension L. What is the only logical dimension to use for this purpose? Explain. Is it necessary to use this dimension in a Froude number if similarity between two systems with complete geometric similarity is under consideration? Explain.

2. Consider the problem of building a model to study the effect of soil properties on distribution of soil water under furrows used in irrigation. Would complete geometric similarity be necessary? Explain.

3. Explain why the criteria of similarity used by Brooks and Corey for drainage situations would have to be modified for cases involving imbibition into dry soil.

4. Consider the distribution of water in a homogeneous soil profile which has drained to a static condition with a water table at some depth. Describe a simple laboratory system that would be effective as a model for the prototype system. Develop similarity criteria for the variables h and S from the equation of statics and the Brooks-Corey relationship for $S(p_c)$. Show what criteria could be eliminated if the variables are $\hat{h}$ and S_e.

5. Consider a case of 1-dimensional steady flow of liquid through small-diameter horizontal tubes of soil that contain a static interconnected gas phase. Derive the least restrictive similarity criteria for a set of variables that will describe flux rate and the capillary pressure distribution.

6. Derive the least restrictive similarity criteria (for tubes of soil as described in problem 5) if the wetting phase is static and the gas is flowing.

7. What is the purpose of the parameter ϕ in the scale factor for p_c used by petroleum scientists. Does this parameter in the J(S) function make the similarity criteria less restrictive if it is required (as proposed by Leverett et al.) that ϕ be equal? Does it make the criteria of Rapoport less restrictive? Explain.

INDEX